Other McGraw-Hill Titles of Interest

Documenting the Software Development Process: A Handbook of
Structured Techniques by *Ayer/Patrinostro*

Designing, Writing, and Producing Computer Documentation
by *Denton/Kelly*

TQM for Computer Software—2nd Edition by *Dunn/Ullman*

Applied Software Measurement: Assuring Productivity and Quality
by *Jones*

Juran's Quality Control Handbook—4th Edition by *Juran/Gryna*

Quality Planning and Analysis by *Juran/Gryna*

Manager's Guide to Software Engineering by *Pressman*

Quality Control Systems: Procedures for Planning Quality Programs
by *Juran/Gryna*

Quality Engineering in Production Systems by *Taguchi*

Bulletproof Docume

Bulletproof Documentation

Creating Quality through Testing

Dorothy Cady

McGraw-Hill

New York San Francisco Washington, D.C. Auckland Bogotá
Caracas Lisbon London Madrid Mexico City Milan
Montreal New Delhi San Juan Singapore
Sydney Tokyo Toronto

McGraw-Hill

A Division of The McGraw-Hill Companies

hc   2 3 4 5 6 7 8 9 DOC/DOC 9 0 0 9 8 7 6 5

Library of Congress Cataloging-in-Publication Data
Cady, Dorothy L., 1953–
   Bulletproof documentation : creating quality through testing / by
Dorothy Cady.
     p. cm.
   Includes index.
   ISBN 0-07-009631-7
   1. Electronic data processing documentation. I. Title.
QA76.9.D6C33 1995
808'.0666—dc20   95-24626
                                           CIP

Acquisitions editor: Jennifer Holt DiGiovanna
Editorial team: Laura J. Bader, Editor
               Susan W. Kagey, Supervising Editor
               Joanne Slike, Executive Editor
               Joann Woy, Indexer
Production team: Katherine G. Brown, Director
                Susan E. Hansford, Coding
                Toya B. Warner, Computer Artist
                Wanda S. Ditch, Desktop Operator
                Lorie L. White, Proofreading
Design team: Jaclyn J. Boone, Designer
            Katherine Lukaszewicz, Associate Designer

0096317
GEN1

Acknowledgments

Many people contribute to the creation of a book, even though they might not always realize it. Thanking everyone individually is almost impossible, so for all of you who helped me in this endeavor, let me say thank you. I would also like to say thanks to several people individually including

Jennifer Holt DiGiovanna for having faith in me.

Drew Heywood for being the first to have faith in me, and for helping me to make sure I did not let him down.

Keith Newman for supporting me all along and furiously reading my manuscript to keep me from going too far astray.

My family—Raymond, Shana, and Ray—who almost never complained about the long hours I kept.

I would also like to thank those of you who purchased this book. You too have demonstrated faith in me, and I thank you for that. I also want to wish you good luck with your documentation testing efforts. May God keep and guide each and everyone of you.

Contents

4 Conducting documentation tests   *131*

7 Starting a documentation testing group in your company   *223*

Introduction

Though it might not be immediately obvious to those involved in developing documentation, poor quality documentation costs both users and developers a lot of time and money. When documentation is published and it contains errors or is inadequate for the job for which it was designed, companies pay for it. Companies often end up spending additional time and money providing increased after-sales support, in writing, editing, publishing, and distributing documentation revisions, and in some cases, in defending lawsuits brought on as the result of documentation errors. As both a writer and a user of technical and end-user documentation, I have first-hand experience of the cost of poor quality documentation.

A few weeks ago, I purchased a do-it-yourself kit for a children's play structure. It included all of the wood and hardware needed to construct a tower platform, swings, a rope ladder, a slide, and a host of other fun features. After dragging it home and into the backyard, I opened the box only to find that there were no instructions for assembling it. Several long-distance phone calls and a very-helpful customer service department employee later resulted in the facsimile delivery of a 16-page set of instructions. There was just one slight problem, however. As I started to inventory the pieces for assembly, I discovered that the instructions were designed to be used with more than one version of the play structure. The instructions were written very generically to accommodate the differences between models. The ultimate result was a set of instructions that were almost useless.

The company probably saved money when it wrote these instructions—shipping this one set with several different models of the same product. However, what they saved, I lost. In addition to taking five days instead of one to assemble this play structure, I also had to make three trips to the hardware store to purchase little pieces of hardware needed to correct mistakes made because of the generic instructions.

The company lost too, though they will probably never know exactly how much. My brother-in-law saw the finished play structure and decided to build one for his children as well. However, when I told him what had happened when building this one, he chose to build his from scratch instead of from a kit. The company lost the sale of a kit to my brother-in-law, and probably many more sales as well, as word-of-mouth advertising is a proven and powerful selling (or not selling) tool.

Besides the loss of potential product sales to others, the company's poor quality documentation resulted in additional expenses to them for customer support—office overhead, personnel, and facsimile charges. In addition, not only will I not recommend their product to others, I will not purchase other products made by this company.

The company could have avoided my dissatisfaction, the additional expenses associated with supporting their product, and the loss of future sales if they had tested their documentation before shipping it with their product. If the company had checked to make sure that the quality of their documentation was as good as or better than the quality of their product, not only would they have saved money on this sale, but they would have made money on additional sales as well.

Unfortunately, few companies know where to begin when it comes to improving the quality of their documentation. Many companies simply rely on the only methods previously available to them—having editors, engineers, lawyers, and other in-house personnel review the documentation. But simply reading through a document is rarely enough to ensure its quality. It takes more, sometimes much more. One method for ensuring documentation quality that is available to companies is that of documentation testing. No longer must a company rely solely on the opinions of a couple of writers or engineers. Now it is possible for a company to prove that their documentation meets a high standard for quality.

The company from whom I purchased that children's play structure could have alleviated many of the problems that I had with their documentation, if they had put it through the right kind of documentation test before releasing it with their product. If the company had a documentation testing group to test their documentation, they would have had several test options available to them, all of which you will learn about as you read this book.

If this company had experienced documentation testers on staff, they would have known that the type of documentation they were producing required an integration level documentation test to ensure that it met its user's needs. For this company's documentation, an integration test would have included testing the document's instructions against each type of play structure they were shipping. Such a test would have shown them that these generic instructions were, at best, difficult to follow.

Had they conducted such a test on their documentation, that company might have discovered ways to improve their document, as well as places in the document to put exception information for different play structures. Or, they might have decided that it would be best to write and print different sets of instructions for each type of play structure. Regardless of their final decision, at least they would have had some practical input from which to decide, and then possibly revise their documentation. Without any testing performed on their documentation to improve its quality, this company will continue to pay for the negative experiences of its customers.

To be fair, not all companies have poor quality documentation, even though they do not have a documentation testing group. Some companies have implemented methods of quality control, such as usability testing.

Intuit, Inc., is a software manufacturer that produces accounting and tax applications for personal and home use. This company has successfully implemented usability testing in their product development process to produce software often described as being superbly easy to use. Intuit employees spend time checking their product's interface designs with groups of users who represent potential buyers. Intuit has successfully improved their product design and related documentation by taking seriously the implementation and results of product usability testing.

Usability testing is one level of documentation testing, but is often used by many companies to the exclusion of any other types of tests. As the success of Intuit and other companies has shown, usability testing has its place. It can help a company to see how easy or difficult users find their product, and to understand user preferences. But it is not a cure-all.

For example, in the case of the children's play structure, a usability test might have helped the company determine which color shade cloth was generally preferred. It might even have helped the company see whether or not the product's potential buyers could understand the instruction diagrams. But, by its nature, a usability test could not have told them that the documentation, start to finish, would be more difficult to use with some types of play structures than with others.

Usability testing, as a type of documentation test, might have helped to somewhat improve the company's written instructions. It might even have helped to make the graphics, and thus the overall appearance of the document, a little bit better. But no matter how good their documentation appeared, it still needed improvement, as does most documentation. There are very few written documents that cannot be improved, and documentation testing is one of the most effective methods for improving the overall quality of a company's documentation.

As a senior systems analyst for a retail department store chain, it was my assignment to design, develop, and document a computerized internal accounts receivable and inventory ordering system. This system was responsible for tracking sales by product, comparing sales against the inventory database, updating the inventory database with purchase receipts, generating product purchase orders, and paying purchase invoices.

To ensure both the quality and effectiveness of the system, as well as the quality and effectiveness of the procedure manual, a co-worker and I conducted a documentation test in conjunction with a staff training session. We provided the software, the computers, and individual copies of the procedure manual to each of twelve newly hired accounts receivable clerks.

For one full week we walked through each procedure in the manual. In some cases, it was necessary for us to explain a process or modify a procedure. When we encountered these instances, we used feedback and input from the accounts receivable clerks, making notes and later changing the procedure manual. When the test was complete and the corrections were entered, the result was a high-quality procedure manual that could be used by any new hire to learn the responsibilities and procedures of their job. New hires no longer needed to have an individual mentor or supervisor looking over their shoulder to help them learn their job.

A test as extensive as this one is certainly not necessary in all cases. In this case, however, it was necessary because we were testing not only the documentation, but an entirely new accounts receivable system. The end result was, however, worth the effort to this company, because of both the time and money that it later saved the company. Developing this system and associated documentation, and then conducting this documentation test to ensure the quality of the procedure manual, made it possible for the company to save thousands of dollars implementing the system, and thousands more in training and hiring costs. It even made it possible for the company to introduce a new, lower-paying, accounts receivable clerical position that did not require either a degree in accounting or experience with computerized accounting software.

Though your company might not experience results as drastic as this company did when it used documentation testing to improve its accounts receivable procedures manual, implementing the suggestions provided by this book can help your company benefit from improved document quality as a result of documentation testing.

This book is intended to encourage individuals and companies to improve the quality of their documentation by implementing documentation testing. It explains documentation testing, and shows you what you need to know and to do in order to implement documentation testing in your own company.

Who should read this book

If you write or edit technical or end-user documentation, then reading this book will show you how to produce higher-quality documentation than you can produce without documentation testing, and in about the same amount of time. Reading this book and implementing its suggestions and procedures will help to contribute to your company's bottom-line profit by reducing some of the costs of production and maintenance associated with poor quality or inaccurate documentation.

Reading and following the suggestions in this book can help to reduce the number and frequency of document revisions, and the number and frequency of documentation errata. By testing documentation before it is published and distributed, important information that might be missing can be found and added now instead of at a much higher cost later. By catching and correcting mistakes that would otherwise require a rewrite or update of the documentation, you can lower writing, editing, printing, and distribution costs.

Reading and following the suggestions in this book can also help to reduce the number of pages in the document. Fewer document pages means reduced printing and distribution costs. A documentation tester can find and eliminate duplicate information or information that is restated using different words.

Using the information in this book can also help to substantially reduce the quantity and length of after-sale service and support calls, resulting in increased company profitability. Rarely can you completely eliminate after-sale support and service costs. However, the fewer service and support calls you receive, the more money you make on each sale.

If you are a software or other product engineer or developer, you should read this book to learn about documentation testing, whether or not you have to write user documentation for the products you develop. If you do have to write your own documentation, understanding and implementing documentation testing can point out areas of your documentation that are too technical or too confusing for your audience. Understanding what documentation testing is and how to use it can substantially improve the documentation that you write. If someone else writes the documentation for you, understanding the documentation process and how and where you fit in can make the entire process of documentation production easier for you.

If you are a software or other product tester, you should read this book to understand how you can help to increase the quality of the documentation for the products you test. Finding documentation errors is often a natural result of product testing, particularly in software testing. Taking a more professional approach to testing documentation along with the product can be a very cost-effective method of improving overall product quality.

If you are a supervisor or manager, you should read this book to help you understand the process and benefits of implementing documentation testing in your company. You should also read this book to understand how documentation testing can help your company protect itself from the potential legal ramifications of inaccurate documentation—bad product publicity and lawsuits from users who suffer "damage" as a result of inaccurate product documentation.

No matter what your relationship to documentation—writer, editor, engineer, tester, developer, or manager—understanding and implementing some or all of the techniques and suggestions in this book can substantially improve your company's documentation. Improved documentation means reduced development, production, and after-market costs, and increased profitability.

What you will learn from this book

After reading this book, you will understand

- The documentation testing process—what it is and how to conduct the different tests—as well as the benefits and drawbacks of implementing documentation testing in your company
- What a documentation tester is responsible for, how a tester can improve the quality of your documentation, and what it takes to be a successful documentation tester

- How documentation testing fits into both the product development and documentation development cycles
- How to prepare for, determine the appropriate level of, schedule, successfully conduct, and effectively document and prepare written reports on different types of documentation tests
- What is involved in starting a documentation testing group in your own company

What this book covers

Bulletproof Documentation is divided into 10 chapters. Each chapter covers a specific topic related to documentation testing. The layout is designed for the individual who knows little or nothing about documentation testing. It begins by introducing you to documentation testing, and takes you through the process of implementing documentation testing in your own company.

Chapter 1 discusses the function and purpose of documentation testing, defining document quality as one of the most important reasons to conduct documentation testing. It then discusses why quality documentation is important, and the benefits and drawbacks of documentation testing. Finally, this chapter provides an overview of the documentation testing process to help define documentation testing, as well as to give you a feel for what you can expect to learn from this book about methods of testing and types of documentation tests.

Chapter 2 discusses how documentation testing relates to both the product and document development processes. It provides an overview of both, and explains where and how documentation testing fits into these processes.

Chapter 3 shows how a documentation tester prepares to conduct a documentation test. It covers topics such as why a tester should attend product and documentation team meetings, what types of documentation are candidates for testing, what types of documentation tests can be performed, how to choose the most appropriate test for the document to be tested, how to develop and write a documentation testing plan, and what preparations must be made before a documentation test can be conducted.

Chapter 4 explains the process of documentation testing beginning with some basics to keep in mind when conducting a test. It then moves on to end with a discussion of methods for tracking and recording testing results. In between, this chapter explains the differences between testing electronic documentation and testing hard copy (paper) documentation. In addition, it discusses how to conduct each of five basic types of documentation tests including read-through, engineering review, usability, basic functionality, and integration tests. Along with an overview and details of the descriptions and processes to follow for each type of documentation test, this chapter also includes samples of documents on which these different types of tests were conducted.

Chapter 5 discusses how to report testing results so that they will be clearly understood and accepted. It includes discussions of both the informal (verbal) and formal (written) methods of reporting testing results. It also points out how to report product problems discovered while conducting the test. Chapter 5 then expands on the preparation of a written report by outlining and discussing the types of information that should be included as a minimum in most formal testing result reports. This chapter then offers guidelines for presenting negative report findings to help reduce the possibility of alienating or angering the writer and others involved in the product and/or documentation development process. Finally, this chapter explains what a tester should do to follow up after the test results have been presented.

After reading through Chapter 5, you should understand what documentation testing is, how it is conducted, and how it can benefit your company. If you are then interested in obtaining information on the qualifications required for becoming a successful documentation tester, Chapter 6 covers such topics as the responsibilities of a documentation tester, and personality traits, skills, and abilities that contribute to success as a documentation tester. It then provides information about some of the education and training alternatives available to anyone who wants to become a documentation tester.

Chapter 7 provides basic information about the most important things you need to know in order to be successful when starting a documentation testing team in your own company. It explains the types of problems you can expect to encounter, and offers guidance for overcoming these problems. It then discusses how to develop a proposal that will help you sell the idea of starting a documentation testing team in your company. And finally, to help ensure that the proposal gets off on the right foot, this chapter offers basic information on how to present the proposal to management.

To help ensure that qualified testers are hired, Chapter 8 provides information about how to successfully advertise for, interview, and hire documentation testers. It explains what is involved in developing a documentation testing budget, an integral part of establishing any company's documentation testing team. It then continues with information on how to determine skill levels and set pay ranges for documentation testers. Finally, it provides basic guidelines on how to find and hire qualified candidates to fill your documentation tester positions.

Once the documentation testing team exists and is functioning, the information in Chapter 9 will help to keep the documentation testing team moving in the right direction. It provides information relevant to keeping the testing team motivated, and shows one method that is useful for helping individual team members get any additional training they might need in order to continue to grow their skills and subsequently their value to the company.

Chapter 10 offers some final advice in the form of guidelines to follow. It provides suggestions to help avoid the mistakes commonly made by documentation testers, ideas about ways to keep the documentation testing team focused on its goals, advice on how to deal with other company employees who want to dictate how and when to conduct documentation tests, suggestions for maximizing team member talents and abilities, and methods for obtaining company recognition of the documentation testing team and its contributions to product and document quality.

Throughout this book there is information that is intended to help you understand why and how to implement documentation testing in your company. By learning the techniques and following the suggestions that this book provides, you will be able to develop your own, as well as others, documentation quality control skills. In the long run, this will enable your company to produce better written and electronic documentation by implementing the techniques discussed in this book. *Bulletproof Documentation* will not only help you do a better job of meeting the needs of your marketplace, but it will also help prepare your company to meet and beat its product's competition.

When you finish reading this book, you will know how to conduct several types of documentation tests. You will also know what training and experience you will need to become a successful documentation tester. You will be able to show management how documentation testing can improve the quality of its product documentation, reduce both short-term and long-term product costs, and increase your company's net profit as a result. Once you have shown management the benefits of documentation testing, you can then use the information in this book to help set up and run an effective documentation testing group in your company. After all, as a pioneer of documentation testing in your company, who would be a better candidate for that task— *and promotion?!*—than you?

1

An introduction to documentation testing

The goal of anyone who writes, edits, tests, or otherwise produces publishable documentation is to create a perfect document: "Winning publications are carefully planned, well written, precisely edited, gloriously illustrated, beautifully composed, and printed with the most tender of care" (Caernarven-Smith 1992, 137). This definition is not quite complete, however, because it does not include the term *painstakingly tested*, a very important part of the documentation creation process.

It used to be common for a writer to think that his work was finished once the document had been written, edited, and prepared for production. Until the advent of document quality control, that may well have been the case. However, more documentation developers are looking at ways to implement a quality control process into their development process.

One documentation quality control process that is receiving more attention is called *usability testing*. More and more documentation development teams are adding usability testing to their documentation and product development process.

Usability testing is a research tool designed to find and correct deficiencies in products—primarily computer-based and other electronic products—and the documentation that accompanies those products. The intent of usability testing is to ensure that both the product and its documentation are easy to use, easy to learn, satisfying to use, and functionally valuable to the target audience (those who are expected to purchase and use the product and its associated documentation) (Rubin 1994, 26).

Usability testing is a valid approach to researching the usefulness and quality of a particular section of documentation. Even with usability testing added, however, the definition of a winning publication is still incomplete. Usability testing is only one type of documentation test, with limited focus and results. In addition, it is not suitable for all types or sections of documents. Also, usability testing rarely includes the entire document. It is only part of the document quality control process, and therefore is not sufficient as an addition to the definition of a winning publication. Instead, the definition of a winning publication should be revised to read as follows:

Winning publications are carefully planned, well written, precisely edited, gloriously illustrated, beautifully composed, painstakingly tested, and printed with the most tender of care.

To understand the "painstakingly tested" part of the definition of a winning publication, you must first understand documentation testing itself. Understanding documentation testing begins with a knowledge of what it does, where it fits into the development process, and what it accomplishes. Before you can maximize the benefits of implementing documentation testing in your product development environment, you need to have a basic understanding of documentation testing.

This chapter defines documentation testing, explains its relationship to and the importance of quality, and gives you an overview of the documentation testing process. After reading this chapter, you will understand

- The function and purpose of documentation testing
- The importance of quality
- The benefits and drawbacks of documentation testing
- The documentation testing process

The function and purpose of documentation testing

Documentation testing has four primary functions and purposes:

- Gathering information
- Locating and correcting document errors
- Contributing to product development
- Ensuring product and document quality
- Improve produce quality and usability

Gathering information

As with usability testing, documentation testing has as one of its purposes the gathering of information relevant to the usability of the document. Unlike usability testing, however, documentation testing is concerned with gathering information about the quality and usefulness of the document as a whole, not just with how easy and satisfying or valuable portions of it are to use.

Documentation testing helps you to gather information about how easy the product and documentation are to use. It also, however, helps you gather information about other aspects of the product and documentation as well, including

- Areas of the document where factual information is incorrect and must be changed
- Improvements that should be made to the product or the documentation to make it easier to use or improve its value, but which might have to be done at a later date
- Product features that might prove to be troublesome for the users
- Documentation text that can be eliminated, thus reducing production, printing, and distribution costs
- Elements of the documentation that might function more effectively if they are presented electronically instead of in printed documentation, or vice versa
- Information that is relevant, important, and missing from the documentation

In addition to gathering information that can be useful to product designers, developers, testers, and documentation writers, the knowledge that the tester gains of the product itself is very useful to the tester when it comes time to conduct the actual documentation test. A strong product knowledge makes it easier for the tester to determine which type of documentation test is most appropriate for the document. It also helps the tester know where to be particularly diligent when looking for problems in the document.

There are other benefits associated with the tester's information gathering. The documentation tester's product knowledge is often more complete than any other single team member. Because the documentation tester is involved with the product from the start, and must be fully aware of all aspects of the product, the tester often has more exposure to the product than other individuals.

Unlike documentation testers, documentation writers might not be brought heavily into the product until the prototyping phase or later. The writer does not always have the advantage of seeing or influencing the design and development of the product from its early stages as does the documentation tester.

In addition, product designers might not be as fully involved in the product as the documentation tester. The designers' and developers' level of involvement is directly affected by the size and scope of the product. The larger the product, the more narrow the scope of each designer's or developer's view of the product. To successfully complete documentation tests, the documentation tester needs to be fully aware of all aspects of the product, and benefits greatly from being involved with the product from start to finish. Thus the documentation tester might be one of the few individuals who has worked on the product from its conception to its shipment, gathering detailed information about the product as its development progresses.

That process of information gathering leads to the successful fulfillment of another function and purpose of documentation testing—locating and correcting document errors.

Locating and correcting document errors

Documentation testing is, among other things, a process to follow in order to locate and correct errors in documentation. Of the five functions and purposes of documentation testing, correcting errors in documentation may well be the most important. Consider the following situation.

You are a student majoring in computer science. One of your courses uses a textbook that contains an appendix of programming code that you reference when writing programs. After spending hours writing, testing, and unsuccessfully debugging your first program, your instructor tells you that the problem is not with your code. In fact, the only problem is that there is an error in the book's appendix that caused you to use the wrong information in your program. How would you feel? Would this situation cause you to think twice before purchasing another book by this author? I believe that it would. Not only is your reaction understandable, but it is also natural, and reasonably common.

If the error you encountered in this book had been something less important, such as a misplaced page number or incorrect punctuation, you most likely would have overlooked it and forgotten about it. Users of products and their documentation have the same reaction. They are generally willing to forgive and forget simple mistakes that have little impact on them. However, the opposite is true if the error is significant, as described in this situation. Not only are users unwilling to forgive and forget important mistakes, but those types of errors often elicit the opposite reaction—one of anger, and even vengeance. Therefore, it is very important that your product documentation not contain errors. Documentation testing, if conducted properly and timely, can catch and prevent potentially serious errors.

Documentation testing can also catch and eliminate errors that, while they might not be serious, can be inconvenient or troublesome enough to the users to cause them to question their purchase, not recommend your product, or even tell others of their negative experience with your product. In fact, problems in your documentation can prevent people from buying your product to begin with, as users sometimes review the documentation before making the final decision to buy a product.

Stop for a moment and think back to the last time you purchased a product, any product. If it was a major purchase, such as an automobile or a personal computer (PC), you might have taken the time to look at the documentation that came with the product before you bought it.

Imagine that the brand new and very expensive automobile that you are about to purchase has nothing but numbered buttons on the radio. With no labels or other guides to help you figure out how to preset a radio station, you might not be able to set the radio, despite your best trial-and-error efforts. You are eventually forced to either open the documentation and look for instructions or go without ever setting any specific radio stations.

If you choose to use the documentation, your first step is likely to include a quick scan of the table of contents or index to find a reference to using the radio. If no such reference exists, what will you do next? Your choices are limited.

You could play with the radio buttons again, hoping to figure it out accidentally. You could scan through the documentation, looking for the word radio to give you somewhere to at least start your search. Perhaps you would choose to start reading a particular chapter that seems to be a likely place to find information about the radio. Or maybe you would simply start reading the manual from front to back. No matter which of these methods you choose, however, you will waste a great deal of time trying to find information on how to set radio stations.

If you have not yet purchased this car, you probably will leave the dealership in search of another car. Not only will you not buy this car, but you probably will not buy any car from this dealership.

If you have already purchased the car when you discover how difficult it is to set the radio buttons, you might begin to question whether or not you should have bought this car to begin with. This is true of any product you buy, no matter what it is. If it turns out that the product is difficult to use, and that the documentation is of limited or no help, you are likely to wonder why you ever bought the product in the first place.

How can this problem be solved? Ideally, the company would completely redesign the product so that it did not need documentation. In the case of a car radio for example, the buttons for setting radio stations and choosing the band (AM/FM) could be clearly labeled as such, an option that a test of the documentation might have shown to be a reasonable solution. Documentation testing can help to point out and correct, or at least make up for, product deficiencies or idiosyncrasies.

Like the product itself, the documentation is also subject to errors and deficiencies. The best way to find and subsequently eliminate any errors or deficiencies in the product's documentation is to conduct one or more documentation tests on the product documentation. Documentation testing, like product testing, is a method used to find and fix problems before the reader finds them.

Documentation problems come in any or all of several types, but there are three content-type problems that are considered most common:

- Errors
- Omissions
- Discrepancies

Errors
A documentation error can be a simple mistake or it can be the passing on of misinformation. Mistakes include such things as a typographical error, such as typing the word Bat, when you meant to type the word Tab. For example, Fig. 1-1 shows a section of documentation that provides instructions for installing and setting up a software program. This section of the documentation instructs the user to press the TAB key at one point during the process. (Note: The word TAB appears in boldface to make it easier for you to see.)

Obviously, keyboards do not have BAT keys, but they do have TAB keys. As Fig. 1-1 points out, mixing up the letters on this one word can cause confusion for the reader. In addition, mistakes of a technical nature in your product's documentation give rise to doubts about the product itself.

In Fig. 1-1, the reader can follow the instructions to a point, successfully restarting the computer and running the setup program. The reader then has the option of choosing from an existing workgroup or creating a new workgroup. If the reader decides to choose an existing workgroup, the reader must ". . . press the TAB key to move the selection bar to the list of available workgroups" (Cady, 1994). If the procedure had mistakenly instructed the reader to press the BAT key instead of the TAB key, the user would be unable to proceed at this point. Not

1. After rebooting your PC during the last step of the installation
    program, the setup program runs and automatically brings up the
    Setup welcoming screen. From this screen, choose Continue with
    First Time Setup.
2. When prompted, either choose an existing workgroup to connect
    to or choose the Create New Workgroup option.

    To select an existing workgroup, press the **BAT** key to move the
    selection bar to the list of available workgroups. Then use the
    arrow keys to highlight the existing workgroup that you want to
    join and press Enter. Now, press **TAB** until Join Selected Workgroup
    is highlighted, then press Enter. If prompted, type in the Supervisor
    Password for the workgroup you chose, and press Enter.

    To create a new workgroup, press the up arrow to select Create
    New Workgroup, then press Enter. Type in a new workgroup name
    and press Enter. Then when prompted, choose Accept the above
    and continue.

1-1 *Example of a simple typographical error*

only would the reader question the quality of your documentation, but
he might also question the quality of your product as well.

This type of problem is clearly and simply a typographical error,
but not necessarily one that is easy to detect. It is not likely that your
word processor's spell checking option will catch this type of error,
because both BAT and TAB are valid words, and both words are correctly spelled.

You might catch the mistake yourself when you read through
your documentation. More often than not, however, time constraints
do not permit you to do a thorough read-through of your documentation. Even if you do have enough time to read it through, you probably have been working with this document for so long that you can
no longer see your own mistakes.

Perhaps the editor will catch the mistake and correct it for you. In
order for the editor to catch the error though, the editor must be familiar with the keys on the keyboard.

Granted, this is a pretty obvious error, and deliberately so, but it
does point out how simple mistakes can cause major problems for
your reader.

A documentation test done on this document would quickly
catch this type of error so that the writer can fix it. For example, if the
documentation tester is walking through and performing each of the
listed tasks in this procedure, when the tester gets to the step that tells
him to press the BAT key, it will soon be evident that there is no BAT
key to press. The documentation tester makes a note on the docu-

ment about the lack of a BAT key. The tester can then take it one step further and let the author know which key is the actual one (the TAB key) the reader must press. Again, this is an obvious example; it would be less obvious if the documentation were explaining how to arm a warhead and launch a missile from a nuclear submarine.

Omissions

Omissions can be just as harmful as errors in a document. What would happen if, in the Fig. 1-1 example, the instruction to press the TAB key was left out? It is likely that the reader would end up being able to choose only the first workgroup on the list. If that is not the workgroup that the reader wanted to choose, the reader might soon feel hopelessly and helplessly lost and frustrated.

If the omission occurred in instructions that showed the reader how to arm a warhead of a nuclear missile, and the author missed a step that tells the reader to release the safety, the missile might eventually be launched, but not armed. This slight omission might result in a major disaster. A documentation test of this procedure would find and correct the omission. However, an editor or the author simply editing or reading through the document might never notice this type of omission.

Discrepancies

Discrepancies in the documentation are the third category of documentation problems that can be caught and corrected before a document is released to its intended audience (people for whom the document was designed and written). Discrepancies in documents usually take the form of disagreements or inconsistencies in facts or claims (American Heritage Dictionary).

Again, because it points out the potentially devastating affect of a mistake of any type in the documentation that instructs a reader as to the proper procedure for arming a warhead and firing a missile, consider the following question. What would be the result if, at one location in the documentation, the reader was instructed to flip the red switch to arm the warhead, but later in the procedure and before being instructed to fire the missile, the reader was again instructed to flip the switch to arm the warhead?

An astute reader might realize that a discrepancy exists in the documentation, and that flipping the switch a second time will disarm, not arm, the warhead. However, if the reader is under stress or pressure at the time, the reader might blindly follow the instructions, resulting in a potential disaster.

You might never have to participate in the production of documentation as critical as instructions for arming and launching a nuclear missile. However, that does not mean that errors, omissions, or discrepancies in your documentation will not have a detrimental effect on your reader. As long as you are involved in or responsible for the production of documentation, it is your responsibility to make every effort to ensure the document's quality. Documentation testing is designed to help you catch and prevent errors in your documentation. Ensuring that errors are prevented is a natural result of the third function and purpose of documentation testing—contributing to the product's development.

Contributing to product development

To understand how a documentation tester contributes to the product's development, you need to understand the development process. By becoming involved in the product and documentation design and development as early in the process as possible, documentation testers can provide valuable input. Testers who are already familiar with this or similar products have design and development experience to contribute. In addition, experienced testers often have greater knowledge of the product's audience than those who are designing and developing the product. As a designer/developer, it is easy to become so involved in the product itself that you lose sight of the user's needs. A documentation tester provides the insight of an individual who understands the product but views it from a distance. You have heard the saying that sometimes you are so close to a problem that you cannot see the answer, even though it is right in front of you. The documentation tester can solve that problem by being a member of the development team, contributing as needed to the development process.

In general, the product development process consists of seven primary steps, as shown in Fig. 1-2. These steps include
- Propose
- Analyze
- Design
- Prototype
- Test
- Develop
- Ship

As Fig. 1-2 shows, each of the product development steps are interrelated. In addition, some of the product development steps are performed more than one time, and depend on the successful outcome of one step in order to move on to the next.

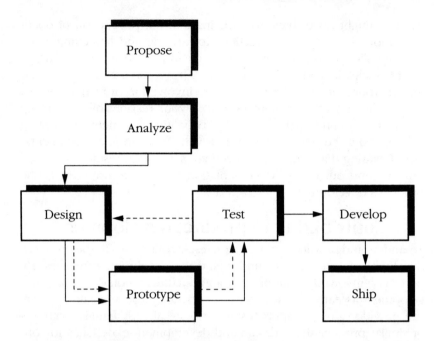

1-2 *Product development process*

For example, once a prototype has been designed and tested, test results might require that the product be redesigned and retested. Each time that testing is not fully successful, the process returns to the design phase for reengineering or redesigning. The product might then be prototyped and tested again, or the process might skip directly to the development phase. In an ideal product development world, prototypes are designed and tested repeatedly until the perfect product is produced. Then, the product is engineered, final testing is conducted, and when the product meets its minimum specifications, the product is built and shipped.

As Fig. 1-3 shows, the product development cycle has inherent within it several areas where documentation can be created. At many of the steps, documentation is internal documentation. That is, the documentation is written for those who are designing, prototyping, developing, and testing the product, not for those who will ultimately use the product—the *end user*. Documentation testing can improve internal documentation as well as end-user documentation, but the main benefits of documentation testing come from testing end-user documentation.

The process of testing a document fits best into the product development cycle once the product is far enough along that writers

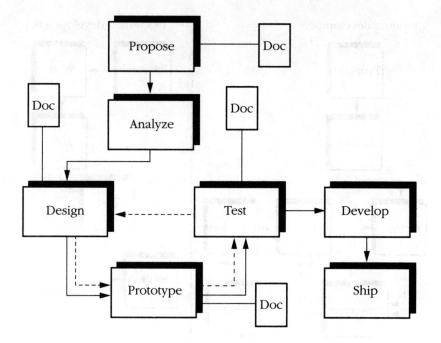

1-3 *Document creation areas within the product development process*

can begin writing the end-user documentation. As Fig. 1-4 shows, this best occurs when the product is in full development. In Fig. 1-4, notice the dotted-line association between documentation testing and product testing. These two testing processes are often conducted simultaneously. But more important is the fact that documentation testers often locate and point out errors, omissions, and discrepancies in the product.

While product and documentation development cycles are not quite as simple or generic as Figs. 1-2, 1-3, and 1-4 demonstrate, these figures do provide a basic idea of where and how documentation testing contributes to the product's development cycle.

Ensuring product and document quality

Documentation testing is also the primary means of ensuring the quality of your product documentation. It is true that most writers and editors add quality to the product documentation by simply being diligent and conscientious when doing their job. However, because of the built-in lead time required to produce product documentation, the documentation itself must often be completed before the product

Product development   Document development

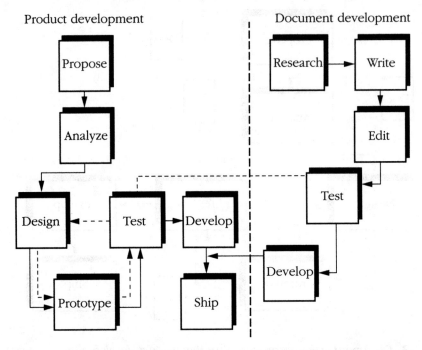

1-4 *Documentation testing within the product development process*

is done. This is particularly true in the software industry, where the documentation might ship both in hard copy and electronic (soft) copy. The lead time required to print and package the documentation can easily be six weeks or more. In the software industry, it is quite possible to fix product problems (also known as software bugs) or make minor modifications to the software during that six-week (or more) time period.

Once the documentation goes to production, even documentation testing cannot change anything in the documentation without seriously impacting the production schedule and budget. However, documentation testing can ensure that what was written up to that point in time is accurate according to the product's development at that time. If stringent development rules are in place and followed to prevent developers from making product changes after the deadline for documentation development, documentation testing does ensure that the documentation correctly represents the finished product. That is, documentation testing goes a long way toward ensuring the quality of both the product and its documentation.

Improving product quality and usability

Most product quality improvements begin with the design of the product's *user interface*—the portion of the product with which the user interacts. In the case of software products, the interface most often appears in the form of menus from which the user makes one or more choices. In the case of a physical product such as a lawn mower, the user interface is varied and might consist of different physical portions of the lawn mower, such as the starter, the handle, or the gas cap. Regardless of what constitutes a product's user interface, however, a documentation tester's involvement in the development of the product and its documentation often helps to improve that interface.

If involved early enough in the development process, a documentation tester can substantially improve the interface, the design, and subsequently the quality of a product and its documentation. By improving the quality of both the product and its documentation, documentation testing contributes to increased product sales, reduced service and support costs, and to company profit.

The importance of quality

My local cable company recently showed a movie about the maiden voyage of the *Titanic*. She was a luxury liner built to such high standards that she was considered absolutely unsinkable. As history has taught us, however, there is no such thing as an unsinkable ship. On her maiden voyage, the *Titanic* struck an iceberg putting a gash in her side so long that it tore through the first four of her watertight compartments. In less than two hours, the *Titanic* sank to the bottom of the ocean taking over 1400 of her passengers with her. Was the Titanic designed and built to high quality standards? It seemed that it was, but if you could ask the passengers that sailed on her maiden voyage, would they agree?

There might be no more ambiguous or relative statement than that which claims that a given item, product, process, or whatever is of the "highest quality." Quality, at any level, is relative to the individual who is concerned about it. That concern, and thus the relativity of quality, is determined by the individual who is creating the product or by the person who will be purchasing or using the product.

This relativity of quality is what makes it so difficult to both define and measure. The difficulty of measuring quality does not make the issue of quality any less important. What it does do, though, is to make it relatively important to establish a definition of quality from which to begin.

Quality defined

There are two primary ways to define product quality. One definition is based on the product's return on investment (ROI). The greater the return, the better the quality is assumed to be. The second but more common way of defining product quality, however, is by measuring the product's value—its usefulness to an individual. Everything else aside, therefore, quality can be defined as sufficient value to some specific person. In other words, quality is relative (Weinberg 1992, 5). This implies that you can both identify the specific person whose opinion of quality matters the most, and that you can subsequently figure out how that person defines quality. When designing and developing a product for resale, this often means that the "average" customer's opinion of product quality is the one that matters. Whether or not this is the correct choice for defining a product's quality is determined by the company's goal.

For example, if the company's goal is to design a product that can be purchased and used by the greatest number of people, then the average customer's opinion of quality might be the most important. If, however, you are developing a product with a more limited sales potential—such as space shuttles—you might have only one or two customers. Meeting the minimum quality requirements of your one or two customers becomes the deciding factor in product quality.

Naturally, there are several factors that go into a person's or a company's decision regarding the quality of any product. Many of these factors are personal, some are political.

In the case of space shuttles, for example, personal factors include such quality issues as whether or not the shuttle can reenter Earth's atmosphere without burning itself and its passengers to a cinder. Political factors determine such quality control issues as the type of fuel that the launch rocket uses based on its effect on Earth's environment.

As you often see played out on the evening news, those factors that are most important to one individual might be the least important factors to another individual. For example, the space shuttles' passengers are far less likely to be concerned about whether or not the fuel used to guide the space shuttle to a smooth and safe reentry pollutes the atmosphere than they are about whether or not it can be safely stored and used on reentry without blowing apart the space shuttle.

Whether the factors that affect an individual's determination of a product's quality are personal or political, there are at least three guidelines that can be commonly applied when assessing a product's quality. They include a product that

- Meets or exceeds user expectations and needs
- Limits product defects
- Improves the product's performance

Meets or exceeds user expectations and needs

To help ensure that a product is of sufficient quality to be marketable and profitable, that product must meet or exceed the user's expectations. "Quality has to be considered from the point of view of the user" (Aguayo 1990, 35). This measure of product quality, however, makes three assumptions.

First, it assumes the users know what they want (see Fig. 1-5), which often is not the case. At the very least, the users must be able to recognize in your product that what they see is what they want. Second, it assumes that you know who will be purchasing your product. That is, it assumes you know and understand your users. Third, it assumes that you know what your users expect to receive from this product, or how they expect to benefit from its purchase and use.

Given these three assumptions, before you can determine whether or not your product meets this level of quality, you must define your po-

1-5 *User indecision affects quality.*

tential product market. Then you must seek to understand what your users expect to receive from your product, or how they expect to be able to use your product.

What the user *wants* to do with a particular product and what the user *needs* to be able to do with a particular product might not be the same. To help ensure the quality of a product, you should measure not only how well it meets your user's needs, but also how well it meets their wants. Then you should try to meet both.

In *Creating Value for Customers*, William A. Band states:

The idea of creating value may, indeed, be reduced to a concept as simple as striving to become ever more "useful" to customers. But, of course, good intentions must be transformed into practical reality. The businesses that will succeed in the decades ahead are not those with advantages defined in terms of internal functions, but those that can become truly market-focused—that is, able to profitably deliver sustainable superior value to their customers. This means being able to do the following:

*   *Choose the target customer and the combination of benefits and price that to the customer would constitute superior value.*
*   *Manage all functions to rigorously reflect this choice of benefits and prices so that the business actually provides and communicates this chosen value and does so at a cost allowing adequate returns.* (Band 1991, 20)

How you define your users and their needs and wants, however, is only half of this quality equation. Choosing which wants or needs you will meet is the second half (Fig. 1-6). The truth is that no matter how hard you try, you might not be able to meet all of the buyer's most common needs and wants and still provide a product at a price the user is willing to pay. There is always a trade-off between how much a user is willing to spend for a product and the quality the user is willing to settle for (Fig. 1-7). The point at which the two meet—cost and quality—also determines how much time and resources your company can spend to design a product with the least number of defects.

Limits product defects

Limiting the number of product defects is another definition for and therefore a potential measurement of product quality. The limit of acceptable defects in a product, and how you measure those defects, also depends on the product, your company's product goals, and the product's intended use.

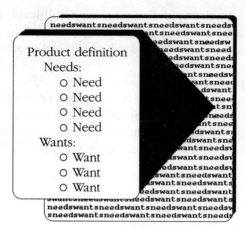

1-6
The buyer's wants and needs affect the final product.

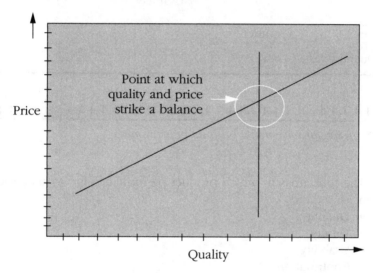

1-7 *Cost plus quality determine time and resources.*

For example, if your company manufactures and sells silk, the quality of that silk can be measured by such factors as its shine, color quality, and thread weight if the final result is the production of bed sheets. Its tensile strength might also be a major quality factor if the silk will eventually be made into parachute material.

In the first case—the manufacturing of bed sheets—the strength of the silk will be less noticeable and less important than the shine and color quality of the silk. In the second case, however, the shine and color quality will be less important than the strength of the silk. Consequently, the type of product defect might be an important measure of quality. Taking care to clearly define what type and level of product defects are acceptable for the product you produce, based on its intended use, becomes an important aspect of product quality.

Improves the product's performance

Improvement in a product's performance can be measured in four areas (see Fig. 1-8). An improvement in one or more of these areas is an indication of improved product quality. The product does not necessarily have to show improvement in all four areas for a product's quality to be improved.

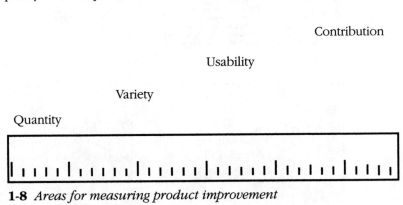

1-8 *Areas for measuring product improvement*

The four areas in which product performance improvement can be measured are

- Quantity
- Variety
- Usability
- Contribution

Quantity might, but usually does not, refer to the amount of the product itself. That is, increasing the number of items of a given product might, but does not necessarily, indicate product quality improvement. For example, if a die cast machine that is capable of producing 1000 die cast toy planes per hour increases its output to 1001 die cast toy planes per hour, this probably does not indicate an improvement in product quality. It might indicate an improvement in the die cast machine itself, but not in the resulting product. If, however, a higher quality of raw material is now being used to make the die cast toy planes, thus resulting in an increase of one toy plane per hour without using additional quantities of material, the quantity increase might be the direct result of an improvement in overall product quality.

The opposite might also be true. Decreasing the amount of product or its associated documentation does not necessarily indicate either an improvement in or a reduction in product quality, although it might indicate one or the other. For example, reducing the quantity of documentation pages in a user manual might indicate improved product quality. That product quality improvement might be in either, or both, of two separate product areas—the product's documentation or the product itself.

The quality of the product's documentation might be seen as improved because the documentation writer was able to reduce the number of pages in the user manual. This reduced page count might be the result of several different factors.

First, changes made to the product by the engineering team might result in a need for fewer pages of documentation. If these changes were made to certain aspects of the product that now make it easier to use, the extra pages of documentation might no longer be needed. The result of these product improvements might be reflected not only in the reduced number of pages of product documentation, but in improvements in the product's actual performance as well.

Second, the product's overall performance improvements might be related to improvements in the product's usability design or its user interface. Such improvements might be less noticeable in the product's overall performance, but more noticeable when these improvements can be more directly related to the product's production costs, including the reduced production costs associated with the product's documentation.

Finally, the documentation writer might simply have found a better way to document various aspects of the product, without any changes having been made to the product itself. These documentation improvements might also result in a reduced total page count for the product's documentation.

The quality of the product's documentation might also be seen as reduced instead of improved when the quantity of documentation pages is reduced. If the page count reduction causes important product information to be eliminated or rendered useless, the documentation's perceived quality is also reduced. If the users of this product believe that the higher the documentation page count the better the product quality, reducing the documentation page count will also reduce the user's perception of the product's quality. Regardless of which factors are responsible for the reduced documentation page count, the results might be an indication of improved product quality, although the opposite might in fact be the case.

There is another reason for a reduction in page count—forced maximum page count. In this instance, reduced page count is not a measure of product quality. If the maximum number of documentation pages has been dictated, this restriction might result in lowering the quality of the documentation, and subsequently of the product.

Variety can also be an indicator of improved quality. In addition, variety can be a catalyst for improving product quality, particularly product documentation quality. Readability and usability of a document are often increased when a conscious effort is made to add variety to a document. Variety can be added to a document in several ways including, but not limited to

- Increasing the number of graphics, photographs, or line drawings
- Using different types of graphics, photographs, or drawings
- Adding color
- Modifying or using a variety of typefaces
- Chunking the information
- Providing additional section headings and subheadings

There are many different ways to add variety to a document. Regardless of which options are used, the act of adding variety might also be a catalyst for, not just an indicator of, improved quality. Being allowed or even encouraged to add variety might also spur product developers and documentation writers to search for and implement other product improvements.

Adding variety, however, does not necessarily mean that the quality of the product documentation has improved. Too much variety, or the wrong type of variety, can actually decrease the quality of the documentation. For example, too many section heads or too many different fonts in a document can make the document difficult to read. Too much variety can also make it more difficult for the reader to find needed information. Too much variety might have the opposite effect, reducing instead of improving quality. Despite its potential drawbacks, however, variety can sometimes be used to measure product quality.

Usability of a product can also be a measure of quality. By first determining how easy or difficult a product is to use, then making product changes and measuring its usability, you also have a measure of the product's quality.

The issue of usability has become an important one in several types of products, but is becoming predominant in the software industry. Early software products were designed, developed, and used by people who spent most of their time working with the product, or at least within the same field of products. They were not designed for just anyone to purchase and use. They often required a specialist.

Many software programs often required several hours of training or trial-and-error experience.

To improve product sales, manufacturers knew that they had to expand their buying public. The only way for software manufacturers to expand their buying public was to make the product inexpensive enough to buy and easy enough for average customers to use. Therefore, one of the measurements of a software product's quality has become its ease of use. Today, many software companies conduct extensive usability tests to ensure that their software products are easy to use.

Usability is a product quality that is easy to measure and quantify. Individuals, after being carefully chosen to reflect the product's intended audience, conduct tests on the product. The goal of usability testing is to determine how easy or difficult it is for users to use a product. The less trouble the user has, the greater the product's usability, and thus, the better the product's quality.

The overall contribution that the product makes can also be an indicator of product quality. There are two particular areas in which the product can make a measurable contribution—company profit and society.

If everything else related to the product remains unchanged, but the product begins to show increased contribution to the company's profit, it might be the result of the quality of the product. Of course, the opposite might be equally true.

Even though the company's profit is not difficult to measure, how much the quality of the product contributed to that profit can be extremely difficult to measure. Sometimes the only way to measure it is by comparing the profitability of two products whose only difference is the quality of the materials used to create those products.

For example, several years ago two condominium developments were built on the same street, side by side, at exactly the same time. The design of the two developments was almost identical. The layout of the condominiums in each development was also the same, as was the sales price. The difference was in the quality of construction materials used to build these two complexes.

At first, all buyers seemed to have received the same value for their money. There were no immediate noticeable quality differences for the buyers. The contractor's immediate profits, however, were significantly different. The contractor who built the lower-quality units initially had a larger net income than did the contractor who built the higher-quality units. After less than one year, however, the quality differences became very apparent, and ultimately negatively impacted the net profit of the developer who had used the lower-quality materials. Failed roofs and other structural damage resulted in class-action lawsuits against the developer who used the low-quality materials. In the end, this developer spent hundreds of thousands of dollars to settle lawsuits, while no such problems or legal actions were initiated against the builder who chose to use higher-quality materials. Over time, the contribution that the quality of materials made to the net income of both developers became very apparent.

This incident can also be used to demonstrate the contribution that a product's quality makes to society. In the case of the lower-quality condominiums, several of the units had to be either partially or completely demolished and rebuilt. The owners were inconvenienced, to say the least. The contribution that the builder of the lower-quality units made to that section of society is questionable.

In addition to a product's quality being defined by its value to a given person or user, there is another way to define product quality— return on investment. As a definition of quality, return on investment is neither personal nor political. It is simply a measure of the cost of developing the product versus the profit the company receives for its investment in the product.

This definition of quality is much easier to measure and to track than some of the other definitions. Sales figures, design and development costs, marketing costs, service and support costs, and other easily identified and tracked items all contribute to the measurement of a product's quality based on this definition—its return on investment. This definition of quality, however, is affected by two factors:

- The timeliness of the return on investment
- The continued usability of the product

The length of time it takes for a company to recover its initial costs associated with the development of the product is an indicator of the product's quality. From a financial standpoint, the product's quality is reduced as the time it takes to recover the product's development costs is increased.

This financial measure of product quality can be compounded or buffered by the second factor—the continued usability of the product. The usable life of a product is also a financial measure of the product's quality. Any product worthy of consideration for production must be capable of sustaining a sales life greater than that needed to recover its initial design and development costs. In other words, the product must continue to be marketable, and consequently to be profitable, after the initial costs of designing and developing the product have been recovered.

Greater usability quality can buffer or offset reduced timeliness quality. A product that might take a very long time to pay back its initial design and development costs can still be considered of high financial quality if it also has a long usability life. A product in this situation might have an overall higher financial quality than a product that takes a short period of time to pay back its initial design and development costs, but which has a shorter usability time.

Why quality is important

Why should we even concern ourselves with the quality of our product? If the market needs or wants the product we sell, will they not still buy the product regardless of its quality? Well, that might be true in a socialistic society or in times of scarce supply, but not in a free-market society or in times of sufficient supply. Product quality is important because of its cost which, according to Philip B. Crosby, is "divided into two areas—the price of nonconformance (PONC) and the price of conformance (POC)" (Crosby 1984, 85). Crosby further states:

> *Prices of nonconformance are all the expenses involved in doing things wrong. This includes the efforts to correct salespersons' orders when they come in, to correct the procedures that are drawn up to implement orders and to correct the product or the service as it goes along, to do work over, and to pay for warranty and other nonconformance claims. When you add all these together it is an enormous amount of money, representing 20 percent or more of sales in manufacturing companies and 35 percent of operating costs in service companies.*
> (Crosby 1984, 85)

Price of conformance is what it is necessary to spend to make things come out right. This includes most of the professional quality functions, all prevention efforts, and quality education. It also covers such areas as procedural or product qualification. It usually represents about 3 to 4 percent of sales in a well-run company.

Using Crosby's figures, it is not difficult to see that the cost of nonconformance can be substantially greater than the cost of conformance. So, why should we concern ourselves with quality, even if users will buy our product despite its flaws? The answer is, of course, profit. Nonconformance can substantially reduce a company's profits.

Profit is not the only reason to be concerned about quality, however, even though many companies must make it their first concern or risk going out of business (see Fig. 1-9). Two other reasons for quality are also important—contributing to society and fulfilling a need.

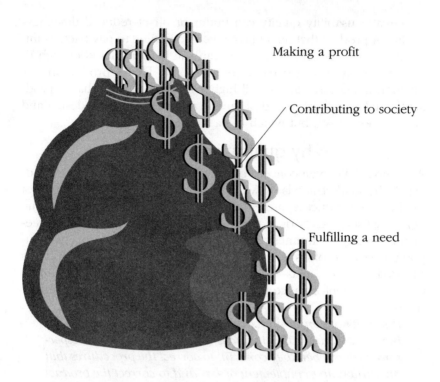

Making a profit

Contributing to society

Fulfilling a need

1-9 *Reasons to be concerned about quality*

The first reason—contributing to society—is an important consideration even if a company is not concerned about the profit they make. For companies that are not profit oriented, their contribution to society is often a very important reason for them to be concerned about quality.

The American Red Cross—a nonprofit organization that provides volunteer help and supplies in times of need—is a good example. They are more concerned with the contribution their organization makes to society rather than with profit. They concentrate on helping people in times of need, often working with victims of natural disasters.

The product the American Red Cross provides is a service rather than a manufactured product, but the principal is still the same. The quality of the service they provide makes a substantial difference in the contribution they make to society. In fact, because the Red Cross' product is primarily a service is a reason why quality is so important. Services are provided by people, and people are one of if not the most critical factor in product quality (Weinberg 1992, 16).

Product quality is also important because the quality of a product often determines how well it fulfills the need it was intended to fulfill. Weinberg makes clear the importance of quality in fulfilling a need:

> *Quality is the ability to consistently get what people need. That means producing what people will value and not producing what people won't value.* (Weinberg 1992, 20)

To successfully fulfill a need and produce a quality product, you must understand the users' needs. If you do so, then when you produce and deliver your product, its quality can be measured by how well it fulfills the needs it was designed to fulfill.

The cost of quality

Making a profit, contributing to society, and fulfilling needs do not come free. As Crosby points out, there are costs associated with designing and building quality into any product, including service products.

Two general categories of product quality costs include

- Obvious costs
- Hidden costs

Obvious costs

In addition to those costs (price of conformance) that Crosby refers to, other obvious costs of product design and development include such things as payroll, raw materials, and general business operating expenses. Costs also include less obvious expenses such as advertising, promotion, and other contract services. When a company increases any of these costs in the name of improving product quality, the increases become tagged as increases in product quality costs.

Once you start looking, you can find other costs that are associated with improving product quality. For example, if your company sends people to special training classes to help them do their individual jobs better, this is a cost of improving product quality. This applies even if the individual taking the training does not directly work on the design and development of the product.

Other obvious costs of improving product quality vary with the type of product. For example, if you are developing an application program such as a spreadsheet, other product quality factors might include such things as

- Software tools purchased for developing and maintaining the product
- Documentation sets and third-party books purchased to further educate those who are developing the product
- Special computer hardware that supports product development based on the type of software tools being used

- Special trainers or consultants hired to assist with the product's development because of their special expertise

Even if you can identify all of the obvious costs of product development, you might not identify all of the costs associated with producing a product. Product development carries with it some hidden costs as well.

Hidden costs

The hidden costs of developing a quality product are sometimes overlooked because they can be difficult to identify and quantify. One of the most important of the hidden costs is associated with the resistance that exists to improving product quality, because it generally means change.

By nature, people resist change. They prefer to be comfortable in their environment, tending to choose patterns and habits with which they are familiar over learning and developing new skills and habits. Improving product quality often requires change. Change often causes stress. Stress often causes resistance. Resistance often increases the cost of improving product quality. Therefore, before you can implement changes to improve product quality, you first have to overcome resistance to change. This is a cost that is not easy to identify, and can, therefore, be a hidden cost of improving product quality.

Another common but sometimes hidden cost of improving product quality is the expense associated with hiring product-related employees. The salary that these people are paid is obvious and easy to measure. Even the amount of money and time spent in such related activities as recruiting, advertising, and processing new hires can be measured. However, these costs are often applied to the company as a cost of doing business, rather than applied to the division or business unit of the company that is producing the product. Therefore, while the new employee's salary and benefits might be charged against the development department, the human resource costs of finding the employee to begin with are generally written off as a cost of doing business and not as an expense or cost of improving product quality.

Not all costs of improving product quality have a monetary value directly associated with them. Stress, discontent, and negative changes in employee moral can also be hidden costs of improving product quality. Unless the employees in your company are truly unconcerned about changes and improvements being made within the company, they are affected by them. It can be nearly impossible to place a monetary value on this type of hidden cost unless it advances so far as to cause work-related stress or other disability claims, or even the loss to the company of that employee. If disability claims or employees leaving the company are the result, at least you can take some measure of the dollar costs associated with these events.

Whether obvious or hidden, conforming or nonconforming, monetary or social, the quality of a product affects the profit of its manufacturer, and of all people and companies associated with it. To ignore quality is to ignore profit—a dangerous gamble in a profit-driven market economy. Improve your odds by improving quality in your company, its documentation, and its products by implementing documentation testing. It can provide you with several benefits, despite its occasional drawbacks.

Benefits and drawbacks of documentation testing

There are drawbacks as well as benefits associated with including documentation testing in your product development environment. Once you learn to deal effectively with the drawbacks, you will also be able to take better advantage of the benefits, of which there are two categories—long-term benefits and short-term benefits.

Long-term benefits

Long-term benefits associated with using documentation testing in your company include
- Better document quality
- Better product quality
- More usable products
- Fewer product releases
- Fewer document rewrites and errata

Taken individually, no one of these benefits provide enough financial incentive to take the plunge into documentation testing. However, taken collectively, these benefits can add to the value, life, and net profit of your company and its products. In addition, as is true with many things, their collective value might far exceed their individual value.

Better document quality

If your primary responsibility is product documentation, you probably make every effort to produce the highest-quality documentation, on time and within budget, that you can. No doubt you have made every effort to hire the most qualified writers and editors (if hiring is your responsibility), effectively train these employees, and implement common quality-control approaches. If you have not yet pursued the implementation of documentation testing, however, you might not have done everything you can do.

Screening to hire qualified writers and editors, then training them to be effective does not guarantee that your final documentation will be of the highest quality. Even the best writers and editors are not perfect. Limited product knowledge, lack of sleep, too much stress, not enough time, and many other factors contribute to the quality of the documentation they, and you, produce.

While writers and editors can be trained to use the products for which they must produce documentation, it is often not possible for them to take the time required to learn all of the aspects of a product. Deadlines, pressure from management, personal needs, and other factors such as the writer's familiarity with this or related products affect the ability of writers and editors to become experts on the products whose documentation they must produce.

For example, it might not be too difficult or time consuming for a writer to learn how to repair a radio if they are responsible for writing a radio repair manual. True, it might seem a little complex at first, but it is likely that because the writer is already familiar with how to use a radio, he or she can readily learn how to repair one. Would it be as easy, however, for a writer to learn how to repair a moon rover?

Because of budget and time constraints, the complexity of the product, and any number of other factors, writers might not have enough time to learn everything they need to know. Managers often do not have the lead time or funds to hire and train writers and editors to the extent required. These types of problems and constraints can severely impact the quality of documentation. However, a qualified documentation tester, performing the right type of documentation test, can help make up for many of the quality problems associated with these constraints. Because of the testing methods used by documentation testers to catch and correct errors and omissions before the document is published, the result is improved document quality.

Better product quality

Documentation testing can also help to improve the quality of the product, and is particularly effective at improving the quality of computer software products such as applications (WordPerfect, Lotus, dBase, etc.) and operating systems (DOS, Unix, NetWare, etc.).

Software is a prime candidate for documentation testing because of the way in which software products and associated documentation are developed. During the typical development cycle of a software product, engineering code is frequently changed to accommodate problems found and requests made as the product's development continues. Such changes might even be made as late in the development cycle as a few days before the master copies of the software are prepared for duplication. While this factor often contributes to the quality of the product, it just as often contributes to errors and omissions in the accompanying documentation.

For example, as the documentation testing team leader for a leading manufacturer of networking software, I conducted a documentation testing study. The study tracked the number of software bugs as well as the number of documentation problems found during the test. The study showed that between 48 and 52 percent of problems discovered during documentation testing were software bugs, not documentation bugs. In addition, the study also found that between 30 and 40 percent of the software bugs were not found by software testers during their regular software product testing process (see Fig. 1-10). These results indicate that, even though software testers are primarily responsible for ensuring the quality of the product, documentation testers contribute to the quality of the product as well by discovering product problems that have not or might not be found by software product testers.

How is it that a documentation tester can find product problems that are not found by the software or product testers? The testers take different approaches to testing. One concentrates on the documentation, while the other concentrates on the product. Taking different

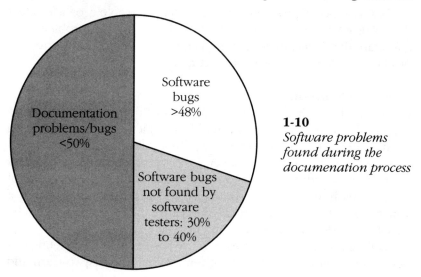

1-10
Software problems found during the documenation process

approaches allows testers to find different types of problems, regardless of the type of product being tested.

The product tester often approaches product testing with more than just a basic understanding of the product and how it works. Frequently, the product tester designs tests based on an anticipated outcome, as well as on a great deal of knowledge about the product. A documentation tester, however, conducts tests based on the contents of a document. The documentation tester might need as much product knowledge as a product tester, but depending on the type of document, can be as successful with only limited product knowledge.

For example, a product tester conducting tests on a radio might need to make certain that the power switch turns the radio on and off. To test the power switch, the product tester flips or presses the power switch twice—once to turn it on and once to turn it off. If the test is successful, the product passes the test.

On the other hand, a documentation tester reads the product documentation and follows its instructions on how to turn the power switch on or off. In the process, the documentation tester might discover a potential product problem. The documentation tester might find that the power switch does turn the radio on and off, but that it is located in a position that causes the user to hit the antenna. A product tester who is simply testing whether or not the power switch turns the radio on and off might never need to extend the antenna. However, if the documentation instructs the user to extend the antenna, then press the power button, the documentation tester will discover the problem.

Is this type of problem a documentation problem or a product design problem? It could be both; but if it is a product problem, it is important that it is caught before it is too late to correct it. Correcting product problems improves product quality.

More usable products

Finding and pointing out a problem such as the one with the radio antenna helps to make a more usable product—one that is easy to learn and easy to use. According to Jeffrey Rubin in his book *Handbook of Usability Testing*, product usability has become a business phenomenon. Whole industries now exist for the sole purpose of teaching us how to use the products we have purchased, many of which "are often initially purchased because they are supposed to be easy to use" (Rubin 1994, xvi).

Selling products as being easy to use has become a marketing tactic—"user friendly" is a marketing term. Why not? Easy-to-learn and easy-to-use products can be an important asset to any company trying to compete in today's global economy. But saying it is easy and making it so are not the same. You must truly concern yourself with the usability of your product if you want to develop a reputation for selling usable products, ultimately helping you to increase your market share.

Documentation testing is one way to increase the usability of your product, before it is too late and too expensive to do so. Documentation testing, implemented at the right time and at the right level in a product's design, can catch problems of usability that even the product testers might not catch. In the long term, this too can result in substantial cost savings, including not having to redesign, reproduce, or rerelease the product.

Fewer product releases

With some products, companies expect to design, produce, and release multiple versions of the same product. In fact, doing so can mean a substantial boost in product sales. This is one reason you so often see "new and improved" printed on various product labels. With some products, however, redesigning, reproducing, and marketing a newer version might have just the opposite effect. In fact, updated releases might be viewed as the company's attempt to fix problems in the original product. Many interim releases of software programs fall into this category.

For some companies, interim releases are a necessary cost of doing business. Subsequently, it stands to reason that if you must issue updates of this nature, getting it as close to right as possible the first time is a major step in keeping overall product costs to a minimum. Finding potential problems with the product before the product goes out the door can be a substantial time and cost saver. Documentation testing is one way to find product problems before a product is ready to be released. In the long term, finding and correcting problems before the product is produced will not only save the company money, but might save the company itself.

Fewer document rewrites and errata

There are different levels of document problems. Errors or omissions in documents that do not result in any serious consequences for the reader might simply be an annoyance to be dealt with at a later date. However, document problems that cause problems for the user might result in serious consequences for the company as well. Documentation testers, because they also find problems in the product, save the company money on fewer releases of the documentation, either as partial or complete rewrites and reprints.

How, if the writer is developing the documentation in direct relationship to the product's development, can the documentation contain errors? In an ideal world, the writer would have stable, locked-down (unchangeable) software, or a fully designed and developed product from which to prepare documentation. The problem is that, in order to meet deadlines and shipping dates, the documentation and the product must often be designed and developed almost simultaneously. While engineers can change software code or a small piece or two of the product almost up to the last minute, the lead time required for producing a product's accompanying documentation is not that generous. The typical lead time for a single book can easily be six to eight weeks. That means the documentation must be ready for production at least six to eight weeks before the engineers stop coding and fixing software, or stop making small product changes. Fixes and changes to the product or to the software code that are relatively simple for the engineer can result in hours of rewriting for the documentation writer.

Most engineers, like most documentation writers, are driven by a desire for a high-quality product. There is a constant temptation to make last-minute changes to improve the product's quality. Granted, even product engineers have production lead times to meet, but even so, last-minute product changes still occur and can have a big impact on the documentation.

While documentation testing cannot and will not prevent product engineers from making product design changes that improve product quality, it does contribute substantially to finding the discrepancies between the product documentation and the actual product. This might also result in another long-term benefit—fewer errata pages.

Fewer errata pages

There might not always be product changes that are important enough or big enough to require a complete rewrite of the documentation. When problems, corrections, or other information must be documented and distributed before the next scheduled printing of the documentation, document errata are necessary. It is a reasonable alternative to redoing the entire set of documentation. Testing the original documentation can improve its quality, resulting in a reduction in the number of errata pages needed later. Better-quality documentation might even eliminate the need to errata your product's documentation. At the very least, testing the errata pages helps to make certain that the changes are also correct.

All of these benefits and their related cost savings can be categorized as long-term benefits. In some instances, they can also provide short-term benefits. There are additional short-term benefits associated with documentation testing as well.

Short-term benefits

Short-term benefits are those benefits that you notice and can take advantage of quickly. Implementing documentation testing in your product development environment often brings with it several short-term benefits including
- Reduced documentation page count
- Improved product design
- Reduced after-sale service costs
- Reduced training costs

Reduced documentation page count
The first short-term benefit that is often the most noticeable is the potential for reducing the number of printed documentation pages. Just as careful design and testing of a software product can help to reduce the number of lines of code, so too can the careful design and testing of documentation help reduce the number of lines of documentation, resulting in an overall reduction in the number of printed pages.

Documentation testing can also point out ways in which the information conveyed in the written documentation can be made more available to the user, at a lower cost. For example, equipment that might be dangerous can be documented as such by including the information in the written documentation. However, unless the user reads the documentation before operating the equipment, it might be far more effective to put a warning label on the equipment itself. A good example of this approach is the warning label found on the power supplies of personal computers (see Fig. 1-11.)

Usability testing is particularly effective in reducing documentation page counts. Usability testing can point out areas in the documentation that the user either cannot or will not read and follow. These areas then become targets for redesigning or rewriting, often resulting in reduced page counts as well as in additional product improvements.

Improved product design
Product improvements can also be a short-term benefit of documentation testing because, while you might experience additional costs in designing the product, the improvements can result in reductions in production or delivery costs.

WARNING

Hazardous voltages
contained within this
power supply; not user
serviceable. Return to
service center for
repair.

1-11 *PC warning label*

Early in my career as a systems analyst, one of my assignments was to review the process of interwarehouse transfers for a national retail department store chain. For these warehouses, their product was a service—the shipping and receiving of merchandise. The main problem the company was trying to solve was excessive product loss. The company was losing merchandise during the transfer from one warehouse location to another. The company's management was convinced the losses were either due to employee theft or mishandling of the merchandise. Whichever was the case, the company felt its monetary investment in a systems analyst was worth it to find and correct the problems.

To correct the problem, however, I had to first find the cause. Step one required determining how merchandise was physically processed and transferred. After visiting three typical warehouse locations and observing the process, I had a pretty good idea of how merchandise was ordered, shipped, and received under the current interwarehouse transfer system.

The next step was to perform a documentation test on their written procedures. Taking what I had learned from visiting these warehouses and testing their written warehouse procedures against actual work flows, I noticed two problems, one of which seemed to be responsible for the majority of the losses. There was one portion of the documentation being followed exactly as outlined, to the detriment instead of the success of the end result and the company.

The documentation required all small or valuable merchandise be stored in a caged, secure area of the warehouse. As instructed, employees placed all such merchandise into this caged area. The cage was generally located in an area of the warehouse that was out of the way of the forklift drivers. As a result, it was generally isolated in the back of the warehouse, out of view. Because merchandise was moved into and out of this caged area frequently, the cage was generally left unlocked throughout the day with no one person assigned to attend it. As a result, merchandise could easily be removed from this area without anyone knowing it. Also, because so many warehouse employees went in and out of this area, it was not easy to point to any one individual who might be responsible for missing items.

To solve the problem, the procedures and documentation were modified to specify several changes. First, the location of the cage was specified in the procedure manual, requiring that it be located where it could be seen from the warehouse manager's office. As warehouse employees rarely needed to move their forklifts around the manager's office, this new location did not interfere with the moving of warehouse merchandise. Second, any requests for merchandise that had to come from this secured area had to be made on a separate request form. Third, only one individual was given these requests, along with complete responsibility for the merchandise going into and out of the cage.

These changes still left the cage open all day, so as not to interfere with inventory processing. The result was a substantial (up to 97 percent in one warehouse) reduction in the loss of this type of merchandise. As a side benefit, the warehouses also noted a reduction in the amount of time it took to retrieve and transfer this merchandise. These three relatively minor changes reduced the interwarehouse transfer time of this merchandise from five to three days, and reduced the loss of merchandise by an average of 90 percent.

Perhaps this change of procedure should have been obvious to someone in the warehouse. However, it is sometimes very difficult to see things about your job that might seem obvious to someone who is less involved in the daily activities. That is one of the reasons why documentation testing is so successful. Regardless of whether documentation testing is being performed on the procedures followed by warehouse personnel or the approach a user takes to working with an application program, documentation testing can help to find and fix product design problems before they become too costly or difficult to correct. The end result can be a reduction in the cost of producing the product, even when that product is a service.

Documentation testing provides two other short-term benefits— the design and development of self-documenting products, and the design and development of user-friendly products. A *self-documenting product* is one in which the product is so straightforward and easy to use that you can either eliminate most of the documentation or make it so minimal as to be able to place it on the product itself. A simple example can be found in the computer hardware industry.

When personal computers first became available, they were sold with instruction manuals that showed you how to assemble them. Later, the major pieces of equipment came already assembled. You still had to plug the keyboard, mouse, printer, monitor, and other devices into the base unit, however, and a small amount of documentation was usually sufficient to show you where and how to plug in these components. Today's PCs are designed so that little pictures (icons) on the back of the base unit show you exactly where to plug in the keyboard or printer (see Fig. 1-12). Thus, the written documentation that used to show you how to perform a task is now eliminated. In other words, it is, at least to some extent, a self-documenting product. Documentation testing contributed to the redesign of computer hardware so as to make the product more self-documenting.

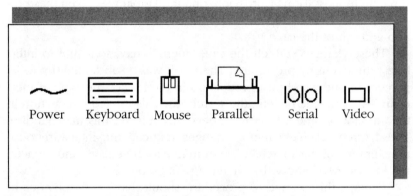

Power   Keyboard   Mouse   Parallel   Serial   Video

1-12 *Icons on PCs show you where to plug in each item.*

By default, such documentation testing can also make the product more user friendly. Again, using the personal computer as an example, today's PCs are far easier for people to set up than ever before because product documentation testing has helped to define areas that needed improvement. Product design improvement also brings with it other important benefits.

Reduced after-sale service costs
Products that ship with problems or deficiencies, known or unknown, often generate calls for help to the company's service and support lines. Documentation errors also generate service and support calls,

and can easily account for 10 percent or more of the calls a company receives. Regardless of how good a product might be, if the documentation provides incorrect, erroneous, unintelligible, or insufficient information or instructions, the user will have difficulty using it. Fixing problems, errors, and technical inaccuracies in the documentation before it goes out the door can be a very cost-effective method of reducing the total number of service and support calls.

As a rule, if your documentation contains typographical errors, the user might notice them but is not likely to call the service and support line to point them out. If the documentation is difficult but not impossible to read and follow, and the user can work around the difficulty, it is also unlikely that the user will call the service and support line for help. However, if the documentation has an error in it, and that error causes problems with the product, you can almost bet the error will result in multiple service and support calls.

Errors in the documentation do not have to be major to generate service and support calls. For example, if you have never used a gas hot water heater before, you might not know that to light the pilot light you must hold down the associated pilot light button for a few seconds once the pilot light flame starts to burn. Failing to hold down this button for a few seconds before releasing it causes the pilot light to go right back out. After you light the pilot light several times, only to have it go right back out again, you are likely to become convinced that there is something wrong with your water heater.

Your first approach might be to read the documentation. If the documentation neglects to include "hold the pilot light button for ten seconds after lighting the pilot light," you will never know what you are doing wrong. Eventually, you will either return the water heater to the store or you will call the company for help. Calling for help is often the easiest, and therefore, the first approach that an already frustrated consumer will take.

Documentation testing could have caught and corrected this flaw in the documentation. Better yet, documentation testing, particularly usability testing, might have pointed out that the instruction should have been placed on the water heater itself, near the pilot light, rather than in a manual the user might read only as a last resort. Such an approach would make this type of service and support call unnecessary, reducing the company's cost of product service and support.

Reduced training costs
Reducing the costs of product training can also be a short-term benefit of documentation testing. While not every company that manu-

facturers a product also uses it themselves, many companies do use their own products. This is particularly true in the software industry. It is very likely that companies such as Microsoft, WordPerfect, and Borland all use their own software products. In instances such as these, in-house product training becomes a big issue and expense. If you design the products that you sell to be easy for your own employees to learn and use, you will be saving time and money for not only your customers, but for your company as well. The easier a product is to use, and the more helpful the documentation is, the simpler and less costly the training becomes.

There is one other short-term benefit derived from using documentation testers in the development of a product—quicker and simpler CBTs (computer-based training programs) and on-line help. Quicker and simpler means reduced up-front costs and increased return on investment in both the short and long term.

How to maximize the benefits of documentation testing

Now that you know what benefits you can gain by using documentation testing, the next most important step would be to look at how you can maximize the benefits of documentation testing in your own company. You have already taken the first step along that road by purchasing and reading this book. Before you can make the most of any product or service, however, you need to understand what it can do for you, how it can harm or help you, and what it is all about. That is just what this book endeavors to do. Once you finish reading this book, you can maximize the benefits of documentation testing for your company by putting into action some or most of its suggestions.

Almost any company that produces documentation can benefit from implementing some form or level of documentation testing. It does not matter whether you are producing application programs or hot water heaters, or simply trying to reduce overhead and improve the efficiency of your warehouse transfer system. Look at the options available to you in your business, find ways to successfully implement those aspects of documentation testing that you believe will bring you the greatest return on your investment, and you will be well on your way to maximizing the benefits of documentation testing for your company.

Of course, nothing in life is free. There are also some drawbacks to documentation testing. Lest you overlook them and suffer all the more because of it, I have pointed out some of the more notorious ones for you.

Drawbacks of documentation testing

There are three main drawbacks associated with implementing documentation testing in your product development environment:

- Potential increase in development costs
- Team interdependence
- Resistance to change

Potential increase in development costs

The first and generally most obvious drawback of adding documentation testing to your product development cycle is the potential for a substantial increase in initial product development costs. Adding documentation testing often requires adding personnel to perform the testing. Increases in staff can quickly increase the cost of developing a product.

Some companies attempt to overcome this particular drawback by simply assigning the responsibility of conducting documentation tests to one or more of their existing employees (see Fig. 1-13.) While this might be an acceptable tactic on very small projects, it has its own drawbacks. Those drawbacks include

- A reduction of the overall productivity of the assigned individual
- A tester with insufficient writing, editing, and testing skills to successfully complete the job and add to the product's value
- A lack of a fresh viewpoint

Reduction in overall productivity Assigning an existing employee to the task of performing documentation testing might reduce the overall productivity of the assigned individual, particularly in their current areas of responsibility. Ultimately, this is usually an approach that works only temporarily, at best, and can increase testing costs in the long run.

The most logical choice for product documentation testing is someone who already understands testing in general. Therefore, a company often gives documentation testing responsibility to an existing product tester. Unless that existing tester has plenty of extra time on his hands, he must figure out how to test both the product and the documentation—usually within the same amount of time given to test just the product. Sometimes the allowed testing time is increased, but most often it is not. Subsequently, the quality of testing of both the product and the documentation suffers.

If the product tester is diligent and dedicated, which is often the case, putting in the extra hours required to complete both types of tests might not seem like much of a problem, at least not at first. The

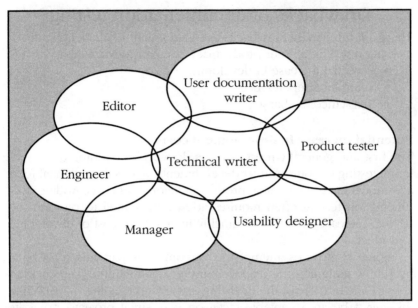

1-13 *Some companies assign existing employees the additional task of testing documents.*

company usually believes that they will only have to do it this once for this one product release, after which the company will make an effort to see that it does not have to be done this way again. If the company pays this tester a salary, which might be one reason why they chose this particular employee to conduct the documentation test, the company will have no objection to the employee working overtime. In fact, the company might expect the employee to work overtime without additional compensation. As long as the company is not paying extra for the employee's overtime, they might feel that they are getting a good deal. However, the opposite is often true.

The result of overworking an employee is often an employee who feels overworked, underpaid, stressed, and unappreciated. The final result will often be the loss of the employee to the company. While that might seem to shift the cost burden of this approach (subsequently hiring and training a new employee) to another company department, it is just a cost shift. The result is still an increase in the overall cost of the product, and of doing business in general.

In addition, even if the employee does not leave the company for a better job, the salary being paid to this type of employee might be higher to begin with than if the company had simply hired a new documentation tester. Hiring another employee at a lower wage to begin with can save the company money in the long run.

Insufficient skills Besides the potential of burning out and losing a valued employee, there is another problem associated with giving the responsibility for testing the documentation to the product tester. The product tester might not have strong enough editing, writing, or other basic language skills. In order for the documentation tester to suggest to the documentation writer an approach that corrects the document's technical deficiencies or problems, the documentation tester needs to have good writing and communication skills. This might not be a skill set that was originally demanded of the product tester. The lack of these skills can jeopardize the accuracy and effectiveness of the documentation testing results, and cause the product tester to be uncomfortable and less effective in all of his assignments, not just that of documentation testing.

Even though it might seem the product will be getting the best of both testing worlds, and that the product should be just that much better because of it, the opposite might be true. The product might suffer from the reduced participation of the product tester as a product tester. To keep from lengthening the product development cycle, it is often necessary to simultaneously conduct product testing and documentation testing. If both tasks are assigned to one individual, one or the other will suffer. While employees might be able to put in the occasional 16-hour day, most people cannot effectively do so for more than a day or two at a time. Productivity, quality, and product and employee safety might all suffer as a result.

No fresh point of view Using one individual presents another problem—the lack of a fresh point of view. When looking at documentation, lack of a fresh point of view often results in not seeing the errors. In the interwarehouse transfer study discussed earlier, the existing warehouse personnel were too close to the real problem to see and resolve it. It took a fresh point of view from another individual to see and solve the problem.

If the drawbacks of using a product tester to test the documentation seem too overwhelming or unavoidable, the next most logical choice for a documentation tester is another writer. In some instances, this can be a good choice, but often is not the best one. The problems associated with this approach become most apparent when the writer is not relieved of other writing duties, but is asked to take on documentation testing duties as an additional assignment. The same problems that exist with using an existing product tester also exist with using an existing writer. The problems are not limited to just increased product development costs, and the burnout and possible loss of a good employee, however. If you use an existing writer as the documentation tester, there will be other problems that have to be overcome as well.

First, if the writer is the same individual who wrote the documentation, they might not be able to see their own mistakes. This is not a negative comment about writers, just a fact of human nature. It is true in fields other than writing as well. Quality assurance (QA) is not effective when implemented by the product's designer or developer, no matter what the product is or who the designer or developer might be.

Methods and procedures for ensuring product quality must be implemented independently of the product's developer. "Regardless of the organizational structure or composition of the QA activity, it must be independent of all activities affecting quality which it must monitor, inspect, or evaluate" (Shillif and Motiska 1992, 19). Without independent quality control, there is no real assurance—relatively unbiased opinion—of a product's true quality.

If the writer is someone other than the individual who wrote the documentation, some benefit can be gained by having this individual conduct a peer review type of documentation test. This individual might not have the technical knowledge of the product that is necessary in order to successfully conduct any of the other types of documentation tests. (The different types of documentation tests that can be performed are discussed later in this book.)

There are other employees that might come to management's mind as potential candidates for performing documentation testing. However, as you can see, there are definite costs and drawbacks associated with using individuals who do not understand the product, the product's defined audience, or the documentation testing process; who are not or cannot be independent in their assessment of the documentation; or who are not dedicated to documentation testing as their full-time job.

In the end, the least expensive method of adding documentation testing to your product development cycle might be the hiring of an independent documentation tester, even though this adds initial cost to the product's development.

Team interdependence

Adding a documentation tester to a product development team might increase the interdependence of team members for product development and quality. Believing that "someone else" is now responsible for, or capable of ensuring quality, other team members might become more dependent on the tester and take less responsibility for building quality into the product. Is this really a drawback of documentation testing? Generally, if you work in an environment where teamwork and production are highly respected and rewarded, then it is an insignificant drawback, if it is a drawback at all. In fact, the opposite situation might be a bigger problem.

If you work in an environment where everyone keeps to themselves and just does their job to the best of their ability, a documentation tester can be perceived as being invasive to privacy and job security, particularly to product engineers and documentation writers. The result is, at best, a resistance to the documentation tester's efforts and, at worst, a deliberate sabotage of the product or the tester's efforts.

The hesitancy of writers, engineers, and even editors to accept a documentation test of their product can be overcome through education, however. Development team members who are educated as to the benefits of documentation testing soon begin to see the process as a way to improve their product, instead of as an impediment to product development and a threat to their job security.

However, even with strong education, the hesitancy is not likely to be easily overcome unless management fully supports the use of documentation testing as a quality control measure. "Above all, the QA organization must have the complete support and confidence of top management" (Shillif and Motiska 1992, 19). Selling management on documentation testing and getting cooperation from other product development individuals might prove to be the biggest challenge you have to face in implementing documentation testing in your product development environment. In a way, lack of cooperation is just a demonstration of the third drawback to implementing documentation testing into your product development environment—resistance to change.

Resistance to change

Many individuals resist any change in their otherwise comfortable environment. In some companies, resistance to change can be a serious obstacle to be overcome. For other companies, change is so commonplace that the addition of documentation testing might fit in as easily as a size 5 foot fits into a size 7 shoe. If you do find it necessary to overcome a certain amount of resistance, there are three steps you can take to help you effectively deal with this resistance.

First, determine the source of resistance. Find out who is resisting the change. You might find that it is only one person, not several, but that the one person has helped to convince others to resist the change as well.

Next, speak with those who seem to be resistant and attempt to find out what is behind this resistance. Perhaps they are concerned about losing their jobs. Maybe they know little about documentation testing and consider it a waste of time and money. Or perhaps they came from a company where quality assurance was either not an issue or was such a big issue that they simply do not care to deal with it at this company. Regardless of who is resisting the change or why they are resisting the change, to conquer your enemies, you must first seek to understand them.

Finally, once you know the cause of the resistance, work to overcome it. If the problem is a fear of losing of their jobs, attempt to show them how documentation testing can make them more instead of less valuable in their jobs. If they do not understand documentation testing or think it a waste of time, show them how effective it can be and how it can save time in the long run. You must find out who is resistant and why, and then make an effort to sway those individuals to your positive point of view.

There are several ways to introduce documentation testing into your company so as to lessen the negative response that you might receive from the changes that it brings. Some approaches to take can include

- Make certain that you involve those who will be affected by the changes in the planning of the changes.
- Be certain that people receive accurate and, as far as possible, complete information about the changes and how they are likely to be affected by them.
- Allow those who are likely to be affected the opportunity to comment freely about the changes, then give them reasonable explanations and answers.
- Start the change process slowly, thus giving people time to adjust to the changes. Implementing the most important or crucial changes first, then refining the process later with the less important changes, can help ease the change process for everyone involved.

Change is not easy for most people. It sometimes causes reductions in employee performance and moral. But change is important and can be beneficial when it is done for good cause, with sufficient time for implementation and evaluation, and not just for change sake.

An overview of the documentation testing process

The process of testing documentation can be described by dividing it into six specific steps or phases. Each phase guides the documentation tester along the path toward quality improvement in both the documentation and the product. Although the improvements associated with the product are a beneficial by-product of conducting documentation tests, they are no less important or valid than the improvements in document quality. However, in the description of the documentation testing process that follows, little reference is made to the product, as the process itself concentrates on the documentation and the improvements and corrections that the testing process makes to the product documentation.

The six major steps or phases a documentation tester follows in the process of conducting a documentation test are

- Participating in product and document development
- Planning and preparing for the documentation test
- Conducting the test
- Reporting the results
- Following up after the test
- Following up after product release

Participating in product and document development

The ultimate goal of documentation testing is to ensure that the document meets or exceeds the user's needs, is free of errors and omissions, and is of the highest possible quality given the time, budget, and other constraints within which it must be developed. To meet this ultimate goal, the documentation tester must have not only a basic understanding of the document's intended user, but an understanding of the development team and environment as well as their goals and prescribed mission—the problem statement from which the development team is to design and implement a solution.

To understand the user, the documentation tester needs to function like or at least become very familiar with the user. To understand the development team's goals and prescribed mission, the documentation tester must function with or become very familiar with the product and documentation development teams. The best way to do this is to become part of those teams, participating in the development of the product and its associated documentation and attending development meetings.

Frederick P. Brooks, Jr., in an article in *IEEE Software* states the following about designing software—a concept which I believe can be generally applied to the development of almost any product or document:

> *The hardest single part of building a software system is deciding precisely what to build. No other part of the conceptual work is as difficult as establishing the detailed technical requirements, including all the interfaces to people, to machines, and to other software systems.* (Brooks 1987)

Much of the definition and design of any product or document is done, or at least approved, by several individuals working together in a team environment. This approach to product and document development frequently requires and results in numerous meetings. By attending these meetings, the documentation tester has the ability to contribute to the definition, development, and refinement of the product or document being developed. The documentation tester is often an effective representative for the end user, bringing to these design and development meetings the user's perspective.

By attending design and development meetings and making himself a part of the development teams, the tester becomes better prepared to determine which type of test will be the most effective to accomplish the documentation testing goals. In addition, the tester is constantly gathering and updating his own knowledge of the product and its documentation.

Attending development meetings also gives the tester the opportunity to become better acquainted with those whose documents he will be testing. Developing an amicable working relationship with the document's designers, editors, and in particular its writers is an integral step in having the tester's suggestions accepted and the problems that were noted during the test corrected without causing animosity or alienating other team members.

Attending development team meetings lets the documentation tester contribute to the development and quality of the product early in its design, prepare documentation tests, and ultimately gain acceptance of the suggestions and changes the tester recommends as an integral part of improving document quality.

Plan and prepare for the documentation test

As the product and document development moves along, the time comes when the tester determines which type of test will be most suitable and effective for the document being tested. Participating in development and design helps to prepare the tester to test the document. Planning and preparing are important steps in completing a successful documentation test.

Part of planning is choosing the most appropriate test to perform, getting acceptance from all concerned as to which test will be done, and successfully fitting that test into the development process. Part of planning also involves knowing what you will need in order to conduct the test.

Preparing involves gathering materials, supplies, equipment, and so on you will need in order to conduct the actual test. It also means setting up a testing space in which to conduct the test, and getting anything else that is required in order to conduct the test.

For example, if you are testing a new version of a software application program, part of planning is knowing what kind of equipment this program is intended to run on. Part of preparing is gathering the equipment you need so that you can successfully conduct the test.

Of course, part of planning and preparing also means organizing for the test according to the type of test you will be conducting. As in the previous example, if you are going to perform only a read-through level test on the software program's documentation because it has received only minor revisions, you probably do not need to have a complete lab of equipment set up. On the other hand, you might want to have access to a computer running the software so that you can make quick checks or verifications when you have questions.

However you approach it, planning and preparing for the actual documentation test is an important responsibility of a documentation tester.

Conduct the documentation test

The most important responsibility that the documentation tester has is, of course, to conduct a successful and effective documentation test. Whether or not the test you run is successful and effective is determined in part by your first two responsibilities—participating in product and document development, and planning and preparing for the test—as well as by the type of test you choose to run.

As this book shows, you have five common types of documentation tests from which to choose:

- Read-through
- Engineering review
- Usability
- Basic functionality
- Integration

The read-through test includes reading the document, or modified portions of it, to ensure its quality and accuracy. Different types of read-through tests are commonly conducted in many companies, although they might not be readily recognized as such. For example, when one writer conducts a peer review of another writer's document, that writer has performed a measured level of a read-through documentation test. The read-through test is commonly applied to documents that are conceptual in design rather than procedural, or have had limited changes made to them since the last release and were tested more thoroughly at that time.

The engineering review documentation test consists of having one or more individuals who are directly involved in the development of the product review the document. An engineering review documentation test can be performed by a documentation tester if they have had sufficient involvement in the product's development. More often than not, however, this type of test is performed by product engineers or product testers, with the documentation tester being responsible for planning for, preparing for, initiating, reporting the results of, and following up on the engineering review. This type of test is most commonly performed on documents that are highly technical in nature and not intended for a general audience. A scientific paper on the latest developments in genetic research would fall into this category, although testers in nonacademic environments are more often expected or called on to test technical documents aimed at a less-technical audience.

A usability test is commonly conducted on the product, as well as on the product's documentation. The tester's responsibilities in this type of test include determining areas of the document that may be problematic, designing user tests to find and correct those areas, gathering individuals who are representative of the intended audience, conducting the test, and analyzing and reporting the results. If a company already has a usability testing group established, the documentation tester's responsibilities might be more limited, but can range from helping to determine areas of the documentation to be tested to attending the tests as an observer, or anything in between.

In the basic functionality test, the documentation tester is frequently working with a type of documentation that instructs the user to perform specific tasks or steps. In this type of test, the primary concern is the accuracy of the procedure being described, as well as its ease of use. This type of test generally requires a great deal of planning and preparation, as well as time spent in the actual test.

One important aspect of the basic functionality test is that it is conducted based on the most common or frequently encountered setup or environment. For example, this is one of the types of documentation testing that the company who sold me the children's play structure (discussed in the Introduction) might have, or should have, conducted on their assembly instructions. If they had done so, the documentation tester would have prepared by obtaining a standard kit and setting up an area in which to assemble it. When it came time to conduct the test, the tester would have walked through the assembly instructions one step at a time, performing each step and noting the results. The tester would have recorded any problems encountered while following the instructions, writing down the correct or most-effective procedure to follow if the instructions were ambiguous or incorrect.

The integration test is one in which the documentation tester attempts to take into consideration every environment or situation in which the document and product could be used and tests the document accordingly. Again, using the example of the children's play structure, the generic assembly instructions for multiple types of kits would have received an integration test if the documentation tester had completed a basic functionality test on the documentation for each type of play structure the instructions were designed to accompany. As you might well guess, this type of documentation test is compounded with each environment or option you add. Not only do you have to make sure that the documentation is correct for two or more separate environments, you must make certain that the instructions for one of those environments does not negatively impact another. Consequently, the more environments you add, the more complex the test becomes.

By knowing the testing options available to you, and by becoming familiar with the user's and developer's needs and responsibilities, you are better prepared to choose the test that will be the most effective. Because the test you choose must be based on the goals, needs, restraints, and deadlines imposed on you and the development teams, you might not always be able to conduct the type of test you believe to be warranted. You might instead have to conduct the test that will be most effective given the situation. There are trade-offs in documentation testing, just as there are in business and life. The secret to being successful as a documentation tester is in developing a knack for balancing those trade-offs.

Report test results

Once you have become familiar with the product and documentation, done your best to influence and guide its successful development, determined and negotiated the best test to conduct, and successfully conducted that test, you must then report the results.

Reports can be written, verbal, or both. The method you choose should be based on which one will give you the most effective result. As with the entire development process, however, there are trade-offs and constraints that you must work within. Unfortunately, you cannot usually get around them.

Follow-up after the test

Once you have reported the results of your test, you will want to follow up to see what has been done about the results. If you have developed a good rapport with other team members (the documentation writer in particular), been tactful in presenting your findings, and been supportive in correcting problems and errors, you will generally get cooperation in return. Your follow-up responsibilities will then be confined to simply making sure that your changes, comments, suggestions, and recommendations are correctly implemented. To do so you must review the results of those changes against the report you originally presented, and correct anything that is not quite right.

Follow-up after product release

Your follow-up responsibilities do not end with reviewing the results of the changes against the results that you reported. Because you are an integral part of the product and documentation teams, you should also be concerned with and involved in a review of the entire development process, once the product ships. If possible, you should also follow up on the success of the product, and in particular of the documentation, in the marketplace. Does the product and the documentation do what it was designed to do? Have users found problems or errors that the team did not catch or correct? Are there any ways to change the process for the better so that the end result is improved next time around? Your input to these and other questions will help to improve the products and documentation that your company produces in the future.

Your participation as a documentation tester before, during, and after the development and release of a product is important to the success of the product and the company. It is also important to your success, a factor that is no doubt important to you as well.

Summary

Producing product documentation is part of the overall function of product development. Applying proven documentation techniques to product documentation before, during, and after the product documentation has been written improves both product and documentation quality. Documentation testing is a quality assurance technique that you can add to your product development cycle to help improve the quality of both the product and its accompanying documentation.

Documentation testing is a continuous activity—a process of locating and correcting errors that is carried out as part of the complete development cycle. Testing of the product's documentation, and participation in the product's development by documentation testers, is a joint and not a sole and separate operation. The documentation tester's responsibilities toward a product begin the day that the first meeting related to the product is held.

Documentation testing contributes to both product and document quality, as well as to the company's return on investment. By improving product and document quality, documentation testing increases the product's monetary value, as well as its contribution to society and its ability to define and meet needs.

Improving product quality is not free, however. There are costs associated with product quality improvements. Some of those costs are obvious, such as the costs of payroll and raw materials. Other costs are hidden, such as those costs associated with implementing change. Some hidden costs cannot be measured as easily in dollars, but might instead be reflected as reductions or negative changes in employee moral and stress.

By understanding the costs and benefits associated with implementing documentation testing as a quality assurance process in your company, you can make an informed choice. Should documentation testing be a quality assurance goal in your company? If your answer is yes, then you can begin to improve the quality of your product and its associated documentation by developing a firm understanding of what documentation testing is, how it fits into the development process, and how you can add it to your development process.

The benefits of adding documentation testing to your product development cycle are many. This chapter pointed out several. You might even find some of your own as you begin learning more about documentation testing and how it can be implemented in your company's development process. If you do choose to implement documentation testing in your company, I believe you will begin to see ways in which documentation testing can help to improve the design of your company's documentation and products long before all of the development is complete.

Consider, however, the drawbacks as well as the benefits of implementing and using documentation testing, so that you will be well prepared to maximize the benefits while minimizing the drawbacks.

References

Aguayo, Rafael. 1990. *Dr. Deming: The American Who Taught the Japanese About Quality*. New York: Carol Publishing Group.

American Heritage Dictionary, 2nd college ed., s.v. "discrepancy."

Brooks, Frederick P., Jr., "No Silver Bullet: Essence and Accidents of Software Engineering," *IEEE Software* (April 1987):10–19.

Band, William A. 1991. *Creating Value for Customers: Designing and Implementing a Total Corporate Strategy*. New York: John Wiley & Sons.

Cady, Dorothy. 1994. *Inside Personal NetWare*. Indianapolis: New Riders Publishing.

Caernarven-Smith, Patricia. 1992. "Cost & Quality: A Balance Sheet for Winners." In *Society for Technical Communication 1992 Proceedings 39th Annual Conference*. Atlanta, Georgia: Society for Technical Communication.

Crosby, Philip B. 1984. *Quality Without Tears: The Art of Hassle-Free Management*. New York: McGraw-Hill.

Rubin, Jeffrey. 1994. Handbook of Usability Testing. New York: John Wiley & Sons.

Shillif, Karl A., and Motiska, Paul J. 1992. *The Team Approach to Quality*. Milwaukee, Wisconsin: ASQC Quality Press.

Weinberg, Gerald M. 1992. *Quality Software Management*, vol. 1, *Systems Thinking*. New York: Dorset House Publishing.

2

Documentation testing and the development process

Before you can test a product's documentation, you need a product and its associated documentation. Products are conceived and developed by individuals—often referred to as entrepreneurs (particularly if they are successful at developing and marketing a product)—or teams of people working for small, medium, or large companies.

The development of any product begins with an idea or a concept. Where, how, or from whom that idea or concept originates is not a topic of this chapter, or even of this book. Instead, this chapter concentrates on the process of developing the product and its associated documentation once the idea or concept has been conceived and accepted.

After reading this chapter, you will understand
- The product development process
- The documentation development process
- How the documentation tester contributes to development

An overview of the product development process

Product development is rarely an easy process. It is also not always a successful process. Bringing a new or revised product to market, and making a profit on it, is a risky and sometimes very costly adventure. According to Don Debelak in his book *How to Bring a Product to Market for Less Than $5000*, only one of every 500 to 1000 people each year who try to market a product are successful. He offers some reasons for this dismal success rate including

- Lack of marketing knowledge
- Limited product market
- Competition from established companies
- Competition from other product creators

In addition to these reasons, Mr. Debelak offers two other reasons for failure that he considers to be the most common reasons—not choosing the right type of product, and making too many mistakes while developing and marketing the product (Debelak 1992).

Assuming that a company successfully overcomes the first four reasons for failure, and chooses the right type of product, the company must still be successful at developing and marketing that product. Since documentation testers are rarely involved in the marketing of a product, marketing is only briefly mentioned in this chapter. Documentation testers do, however, play an important role in the development of a product, as well as in the development of its accompanying documentation, as explained in chapter 1. Therefore, an understanding of the product development process and of the documentation development process is important for success as a documentation tester. Both are covered in this chapter, along with information about where the documentation tester fits into the product and documentation development processes.

Product planning and research

Successful product development begins with successful product planning and research. Therefore, before developing or producing any product, the first step to be taken is that of planning. Before any planning can take place, however, the need for the product must be assessed. Unless there is a need for the product and that need is established before other steps are taken, the product is doomed to failure before it starts. Here is where many product developers encounter several of the problems Mr. Debelak believes will result in product failure—lack of marketing knowledge, limited product market, and not choosing the right type of product.

Product needs assessment usually requires research. This research is often conducted by a company's marketing personnel. They specialize in following the market, knowing how to assess the need and desire for a particular product, and informing the company as to the viability of a given product or idea for a product. If the company is small, such as a sole proprietorship or partnership, or has only a few employees, the marketing research must be done by these few individuals, or contracted from a company specializing in this type of research. It does not matter who does the research, as long as it is done and done competently.

The primary purpose of the research is also the first step in product planning and research—establishing the need for the product. It does not matter whether the research determines that the need for the product is genuine or created, as long as the need for that product is established. Approximately 100 years ago, an educator named Gerald Stanley Lee said, "A man's success in business today depends upon his power of getting people to believe he has something they want" (Griffith 1990).

An excellent example of convincing people that a need exists and then filling it can be seen in a product developed and sold in the 1960s. A company created and marketed a product know as a pet rock. The product itself was literally a rock marketed as the perfect pet—you did not need to feed, water, clean up after, or in any other way care for your pet rock. Yet, you could take advantage of the idea of this rock as being a pet. It became a fashionable fad to give a pet rock as a gift or as a joke. Did people really need to own a pet rock? As a practical matter, probably not. But because it was fun and faddish, it did fill a need for having fun and following the latest fad.

The type of research that was done for this product could have been as simple as asking people how frequently they needed a perfect pet, or by asking them for their description of a perfect pet. The researchers might have performed more traditional research efforts and conducted focus groups or telephone surveys. They might even have approached people who fit the description of their target customer with a clipboard and questionnaire in hand at the local mall. Or they might simply have come up with this idea one day and hired a marketing or research firm to test out the likelihood of sales for such a product. Whatever approach they took, the developer needed some way of determining whether the product would sell before putting thousands of dollars into building, packaging, marketing, and distributing his product. The developer must have conducted some type of needs assessment in order to determine the basic viability of the product.

The method used to determine the need for a product can take many forms and is greatly influenced by the type of product. In the case of a pet rock, the needs assessment might have been conducted by any or none of the methods previously described. Perhaps the needs assessment was conducted by creating a few pet rocks and giving them away to friends and colleagues to see how well they were accepted. This type of approach is known as *prototyping*, and is a reasonable approach for a product that requires minimal design, production, and distribution. With most products, however, prototyping is rarely this easy, and often cannot be done until much later in the development process. But it does serve to show the range of approaches available for conducting needs assessment and related market research.

Though the pet rock might seem a poor or strange example, it also serves to prove that, while the success of any product is related to the product itself, a product's success is more dependent on the initial planning and research that goes into the product. Before a company spends thousands of dollars on the development, marketing, and distribution of a product, it must first determine the need for and potential of that product through market research, without which the company stands to lose everything it has invested.

In addition to the methods discussed previously, there are several other ways to gather information from customers or potential customers about the need for your product. Some of these methods include

- Contacting potential or existing customers using questionnaires
- Holding roundtable discussions with existing or potential customers
- Conducting telephone or electronic mail surveys

Besides customers, there are internal individuals and internal information that can provide your company with some of the information needed to determine the potential of a product. Some of these individuals and types of information include

- Service and support personnel and statistics
- Customer complaint personnel and complaint logs
- Testing personnel and testing reports from earlier versions of this and related products
- Other employees who deal directly with customers, and any related documents or reports prepared by these individuals

Companies have many choices for gathering information about the potential and need for a proposed product, some of which might prove to be very costly. In fact, some product ideas do not always come from research. "In 1957, Vic Barouh's company made carbon paper. One day he saw a secretary use chalk to erase a mistake. This is how the idea for Ko-Rec-Type was born" (Griffith 1990). The need was obvious. The research was minimal. The resulting product was a success for Vic Barouh's company, without the expense and time involved in extensive market research and analysis.

Regardless of how your company approaches its responsibility of information gathering, once the needed information is obtained, product planning and development can be started. The sooner it is started, the better, particularly if your product involves new technology. John Handly, a vice president for AT&T, stated that "If you get to market sooner with new technology, you can charge a premium until the others follow" (Griffith 1990). Such an approach gives you the opportunity to recover your costs associated with research, planning, and product development sooner than those who enter the market as your competitors.

Being the first into the marketplace has another benefit as well. Once your name and product become known, it is more difficult for a competitor to establish a presence in the marketplace, providing your product offers quality for a competitive price. This means that, even when you are first with a new technology, you must not forget that it is the product's quality that helps you keep that presence once competitors start producing their own version of your product. One way to ensure the quality of your product and its documentation is by including documentation testers in the product's development.

Most often, a documentation tester does not become heavily involved in a product's development until its viability in the marketplace has been shown. By that time, the company already has a vested interest in the product's success, and the additional participation of a documentation tester on the product development team is seen as one way of helping to ensure the buyer-oriented development of the product. A buyer-oriented product has a better chance of success when it is introduced into the marketplace. Because of the tester's broad perspective of the market, the product, and how it will be used, the documentation tester can provide some or most of the buyer-oriented perspective to other team members, thus helping to ensure the success of the product.

Initial documents

Market research efforts generally result in some type of written materials from which further product development is then directed, assuming the market research proved the viability of the product. One type of document commonly produced is a marketing research document. This document, which describes the product's overall design from the marketing point of view, can have many different titles. Although often just as applicable in other industries, in the software development industry, this type of document (or group of documents) is often referred to as a marketing *specification* document. Some typical titles for this type of document include

- Marketing requirements document
- Product specification
- Requirements definition
- Problem statement
- Product overview
- Marketing research summary

For the sake of this book and simplicity, however, this type of document will be referred to as an MRD (marketing requirements document), although different industries and even different companies within the same industry call it by any of several different names.

An MRD describes the product, details the market needs that it must meet, and provides a time frame within which it must be produced in order to take advantage of the market's desire for the product. The MRD plays an important role in the development of a product to the stage where a test can be performed on the product's documentation. It provides the documentation tester and other product and documentation team members with information about the product to be designed and produced. The MRD guides all of the initial design and development of the product, and helps all team members to begin thinking about and moving toward that single product. It is the starting point for the product's design and development.

In a perfect world, the MRD is the responsibility of those who conduct the product's needs assessment, generally members of the company's marketing staff, although it might be produced by other related individuals such as the company's communications staff. It is not uncommon, however, to find that the MRD is created by product designers, managers, technical writers, or other involved team members because of a lack of available marketing or communications staff.

Sometimes, products and their related MRDs are created because company management believes there is a need for a particular product, even though formal market research might not be conducted. In this instance, a group of people are directed to produce a product for which no formal MRD or related document exists. Because the team needs somewhere to begin, one or more of the team members might end up producing the MRD, or a similar document, from which the product's design can begin.

A product cannot be successfully designed and developed unless the team knows what that product is supposed to do, who is likely to buy it, what other products are considered to be its leading competition, and so on. Therefore, the MRD needs to contain the following types of information:

- Audience definition or profile
- Target customer description
- Customer needs and wants to be satisfied
- Product competition
- Marketing outline

Both the audience definition or profile and the description of the target customer are important parts of this document. To make sure that all designers, developers, and other product development participants create a product designed for the prescribed group of buyers and users, the MRD should include an audience profile and a description of the target customer.

While closely related and sometimes one in the same, there can be a significant difference between the audience and the target customer. The audience description centers on those who are likely to make the purchasing decision related to the product. The target customer description centers on those who are likely to use the purchased product. For many products, these are the same. For other products, these might be quite different.

For example, the audience for a doghouse might include such potential buyers as pet owners, pet shop owners, and pet supply retailers. The customer for the product—the doghouse—is a dog, not the dog's owner. The product must be designed not only to convince its audience to purchase it, but also to be functional, usable, and comfortable for its target customer—the dog.

The audience profile and target customer description should include information that lets you readily understand the wants and needs of both your audience and your target customer. They should reflect information about the audience and customer that helps you to make important decisions related to the design and development of the product. The exact information it contains is determined, at least in part, by the type of product you are developing.

For example, if you are developing a software program you might need to know such things as

- Which type of operating system the audience and customers most commonly use
- What level of experience they believe they have with using operating system software

- What other software programs they have experience using
- How experienced they are at using each software package
- What computer experience they have had
- Whether or not they have any training or education in computer software

In the case of software programs, the audience and the customer are likely to be the same. If you are developing a doghouse as suggested before, the two will be different.

A primary reason to create and sell a product is that of satisfying customer needs and wants. Thus, information related to those needs and wants is an important part of the MRD. There are different ways to present the information regarding customer wants and needs. One effective way to present this information is by listing the problems or needs of the audience and target customer, followed by a description of how the product meets each need.

Using the doghouse as an example, one such need and how that need will be met is described in Fig. 2-1. As the figure shows, the MRD not only lists an audience and customer need, but also presents one potential solution. Although product developers might seek other means of successfully meeting the need, the information provided by the MRD gives them a starting point from which to work. In addition, and in this case, the potential solution presented also provides the development team with information about competing products.

Product competition includes anything the prospective buyer (audience) might choose to purchase instead of your product. To ensure that yours is the product the audience buys, your product needs to be of better quality for a similar price to that of the competition or offer features that the competition does not offer. Before a product can be designed to be of better quality than the competition, or to have more features than the competition, you must understand the competition. Therefore, information about your product's competition is an important part of the MRD.

Product design can also be greatly influenced by how the product will be marketed. Understanding how the company intends to market the product through the introduction of a marketing outline or guidelines is, therefore, also an important part of the MRD.

Once again using the doghouse as an example, if the product will not be marketed to pet shop owners or pet supply retailers, but will instead be marketed directly to pet owners, the audience with which the designers and developers must be concerned is substantially narrowed. Knowing this, the product's designers and developers can emphasize product features that are important to dog owners, ignor-

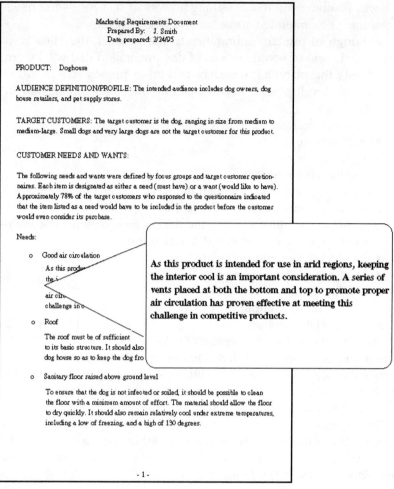

Marketing Requirements Document
Prepared By: J. Smith
Date prepared: 2/24/95

PRODUCT: Doghouse

AUDIENCE DEFINITION/PROFILE: The intended audience includes dog owners, dog house retailers, and pet supply stores.

TARGET CUSTOMERS: The target customer is the dog, ranging in size from medium to medium-large. Small dogs and very large dogs are not the target customer for this product.

CUSTOMER NEEDS AND WANTS:

The following needs and wants were defined by focus groups and target customer questionnaires. Each item is designated as either a need (must have) or a want (would like to have). Approximately 78% of the target customers who responded to the questionnaire indicated that the item listed as a need would have to be included in the product before the customer would even consider its purchase.

Needs:

o   Good air circulation

     As this prod
     the i
     air cir
     challenge in c

o   Roof

     The roof must be of sufficient
     to its basic structure. It should also
     dog house so as to keep the dog fro

o   Sanitary floor raised above ground level

     To ensure that the dog is not infected or soiled, it should be possible to clean the floor with a minimum amount of effort. The material should allow the floor to dry quickly. It should also remain relatively cool under extreme temperatures, including a low of freezing, and a high of 130 degrees.

As this product is intended for use in arid regions, keeping the interior cool is an important consideration. A series of vents placed at both the bottom and top to promote proper air circulation has proven effective at meeting this challenge in competitive products.

- 1 -

2-1 *Marketing requirements document for a doghouse*

ing features that otherwise might be important to only shop owners or pet supply retailers.

The MRD gives the product's designers, developers, and other team members including the documentation tester not only a starting point for product creation, but also detailed information about the intended audience and target customer, the product's expected competition, what product features are needed and wanted by the audience and customers, and how the product is supposed to be marketed. This information makes it possible for the product development team to begin planning for and designing a product that meets the needs and wants of its intended audience and target customer.

Although of primary importance to the product, the MRD is, of course, only one of several pieces of documentation that will be created during the planning, research, and other phases of a product's design and development. Other types of initial documentation include

- Product design requirements
- Architecture
- User interface
- Engineering interface

Once the company has defined the basic product using the MRD, the product's actual design and development requirements must then be established. As with the MRD, the design requirements documentation might have any of several different titles:

- Product design requirements
- Design and development
- Product features

For the sake of simplicity, this document is referred to in this book as the PDR (product design requirements). The PDR provides information about the requirements or needs the product will be designed and developed to fulfill. This list of requirements to be met creates the *scope* of the project. While the MRD specifies what the product *could* contain based on market research and other factors, the PDR specifies what the product *will* contain.

The contents of these two documents—the MRD and the PDR—are not always the same. Sometimes the MRD specifies needs to be met or goals to be accomplished that cannot be implemented for the product at this time. Not being able to meet all of the defined needs does not mean that the product should not be created. The final outcome of many product development cycles is often the result of a balance between what the market says they need and want, and what the producer is capable of delivering.

There is one other important factor that contributes to the final design of the product—the amount of money the buyer is willing to pay for the product. Like many individuals faced with the choices of a wonderful buffet, our eyes are often bigger than our stomachs. That is, buyers often think they want a lot of features and gizmos, but when it comes to paying for them, they are often unwilling to pay as much as it would cost to include all of those features in the product. Therefore, there must be a balance established between the features that can be included and the cost that will be charged for the product. The PDR describes the product as it is feasible to produce it.

The PDR generally includes the following types of information:
- Product strategy
- Product functionality
- Related product requirements
- Marketing versus development requirements

The product strategy section of the PDR begins with a description of what the product is intended to accomplish. That description should be a translation of the operational needs that provided the impetus for the product and what is expected from the final product.

The product strategy section needs more than just a description of the product. It should also include such information as when the product can be ready for release (a projected shipping date), when the product must be ready or risk the possibility that the market for it will no longer exist, what resources will be needed in order to meet the schedule, the value of the product to its intended customer, how the product fits into the company's existing product line (if applicable), and any other related information that might be important to product strategy.

The product functionality section of the PDR provides a list of all requirements that effect the functionality of the product. Along with this list, each need or problem that the product will solve is identified, but without addressing exactly how it will be solved. The details of how each problem or need is solved are addressed in a later document.

When the functionality of the product is addressed, it should include information about several aspects of the product's functionality including such things as product performance standards, usability of the product, level of quality of the product, and reliability of the product. These are general issues that the product functionality section of the PDR should address for most types of products.

In addition to these general issues, the PDR should also address functionality issues specific to the type of product being developed. For example, if the product is a portable ramp for wheelchair access, functionality issues such as how the ramp is to be installed must be addressed. If the product is a software application program, how the product will be manufactured for distribution (CD-ROM, floppy disk, etc.) must be addressed.

In addition to product functionality, the related product requirements section of the PDR should be completed. This section addresses all other areas related to designing and developing the product, even though this section does not specifically address the product itself. For example, how the product is to be built and packaged should be addressed in this section. Other product-related requirements that should be addressed in this section include

- How the product is to be distributed
- What type of documentation will be created for the product and how it is to be distributed
- How product support will be provided, if it will be provided at all
- What training will be needed and how it will be provided for company personnel who will be marketing, selling, or supporting the product
- How the product will be marketed, including such issues as plans for the initial announcement of the product
- Whether or not there are any legal issues such as patents or trademarks to be considered and addressed

The product functionality section of the PDR provides information about all aspects of pre- and postdevelopment of the product not specifically addressed elsewhere in the document.

The last section of the PDR—the marketing versus development requirements section—is intended to ensure that the PDR addresses all of the product requirements listed in the MRD. It can consist of a simple table where the MRD requirements and the PDR's list of product features are compared, or it can be as elaborate as a multipage description of how each product feature fits with the feature list in the PDR. Whatever approach is used, this last section of the PDR is designed to ensure that each need or feature included in the MRD is addressed, either to describe how it will be met or to explain why it is not possible to meet the need or include the feature with the product being developed.

The MRD and PDR provide the designers, developers, and other interested individuals with the basic information about the product, including the needs or wants that it should and can meet. These two documents are part of the planning and research involved in developing a product. They work together to identify the need for and establish the viability of a product.

These two documents are usually created relatively early in the concept phase of a product. At the very least, the information contained in these documents is necessary for a company's management to make an informed decision as to whether or not to proceed with the design and development of the product, and then plan its development. Once the decision is made to proceed with the product's development, and initial planning has been done, the true product development cycle begins.

Product development

The product development cycle is a multiphase process that, at its best, results in the creation and success of a given product. Product development cycles can last anywhere from a few weeks to several years, depending on the product. Regardless of the type of product or how long the development cycle lasts, product development cycles proceed through a series of stages, each one dependent on the successful completion of the one that came before it (Fig. 2-2).

Product development cycle

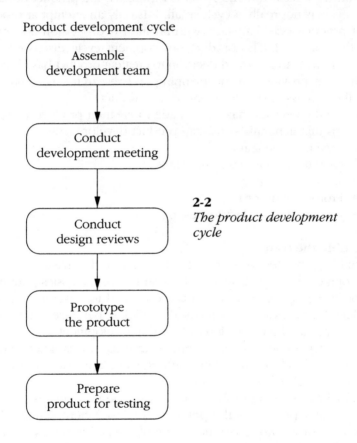

2-2
The product development cycle

The product development cycle described here is not necessarily the only or the most perfect one. It is, however, somewhat typical of product development cycles in general, although product development in your company might proceed somewhat differently from that described in this chapter. Consider the similarities and the differences between the product development cycle described here and your company's product development cycle. Then you will better understand that regardless of how closely the product development cycle in your company does or does not match that described here, the idea of product development as a cycle into which the documentation tester fits is still applicable.

Some companies and individuals believe that the product development cycle includes the planning and research phase previously described, and this is often true. Many product ideas, however, have been put through the planning and research stage, but never make it any further. Unless the end result is a product, the product development cycle is not really a cycle at all. It is only an attempt at producing a product; one that starts with an informed decision and might end there as well. The product development cycle discussed here also starts with an informed decision to create a product based on the information provided in the planning and research stage and ends with the successful development of the product.

Once the decision has been made to create a product, there are five steps that generally constitute product development:

- Assemble the team
- Conduct development meetings
- Conduct design reviews
- Prototype the product
- Prepare the product for testing

Assemble the team

Assembling an effective team is the first step in the five-step product development process. It would also seem to be the easiest, but often is not. In many companies, teams are created to develop a specific product. Members are chosen based on their experience, knowledge, and other related qualifications. Once the development of that particular product is complete, that same team is simply assigned another product to develop. This might be a suitable approach, but carries with it some obstacles to success.

The first potential problem with using the same teams to create product after product is the potential for insufficient variety of experience. For example, assume a team of researchers and writers has successfully developed a set of encyclopedias aimed at college students. The writers and researchers all have experience and training in a variety of subjects such as history, art, geography, and so on. At first glance, this team seems ideally suited to also be assigned the task of creating a set of encyclopedias for grammar school children. But what if none of these team members have any experience with teaching children, or with creating written materials for children of grammar school age? Are they then still the best choice for developing these encyclopedias? Perhaps this same team is still suitable, but might be more so if a child educator or child psychologist were added to the team. Therefore, the first problem associated with creating a team is also the first reason that creating a successful team can be so difficult—team members must be chosen because of their knowledge and experience related to the specific product they will be developing.

Another potential problem can be cooperation among team members. No doubt there have been times in your life when you found that the best way to prevent problems was to avoid them to begin with. If any member of the team does not or cannot get along with one or more of the other team members, that member must either be replaced or risk disharmony among the whole team. As a team member, you might find it possible to get along in the team because you know that once the product is developed, other opportunities to work with other individuals will become available to you. A team member who does not believe this to be an option might feel trapped and resentful.

The simplest solution for the company might be to terminate the troublesome team member. By doing so, however, you might be eliminating one of your most knowledgeable team members. The whole team and the product could suffer as a result. Allowing team members the flexibility to work on different teams, gives them a sense of freedom, as well as a greater depth and breadth of experience. In addition, it lets you combine your best qualified individuals into a unique and completely appropriate team.

Qualifications and the ability to get along with other team members are two important qualities to look at in each potential team member, if you want to create a successful team and product. Variety of experience and approach are other important aspects that should be sought after when creating a development team. One way to accomplish variety is by including individuals who represent different phases of the product's development (Fig. 2-2). Such individuals include writers, engineers, and yes, documentation testers.

Conduct development meetings
Once the team has been assembled, they will start holding development meetings. The initial purpose of these meetings is to create the design of the product. To fulfill this purpose, the team must be organized. The first step to organizing the team is to assign a product manager—someone who is responsible for the successful completion of a product.

The product manager has several initial responsibilities, the first of which is developing a complete understanding of the product to be produced and its intended position in the marketplace. The product manager gets this understanding by reading and questioning the contents of the MRD and individuals who have already been associated with this or similar company products. Without an in-depth understanding, it will be difficult for the product manager to keep the team focused on the intended product—the second responsibility of a product manager.

The product manager might also be responsible for choosing team members, although it is just as common to have members assigned to the team by other company personnel. If the product manager is responsible for choosing team members, then it is also the product manager's responsibility to choose individuals who can contribute not only to the product's development, but who can also contribute new ideas, even if this means a turnover in the makeup of the team. This is particularly important whether you are creating application software, designing a submarine, building a jazz band, etc.

Woody Herman was once asked if the continuing turnover of personnel in his band bothered him. He said the first time some of his star musicians left the band, he was devastated. He thought he would never again have such great players. In time, as personnel changes became a continuing fact of his music world, he said he began to look forward to the newcomers' new ideas and the contributions they made to the ongoing Herd. (Wilson 1992)

Regardless of whether or not the product manager chooses each of the team members or works with a preassigned team, the product manager is still responsible for making certain that the team is complete and that it is effective. This means that if a documentation tester is not assigned to the product team, the product manager should be the one to seek out a documentation tester to join the team.

If someone has not already been assigned to take meeting notes, the product manager is also responsible for ensuring that notes are taken. Different options are available to the product manager. Team members can be asked to take turns or one individual can be assigned to keep the meeting notes.

Who keeps the notes is less important than why they are kept. These notes are needed to be able to answer questions related to decisions that are made. A more important reason, however, is that of being able to show that different approaches to developing the product have been considered, and why it was decided that the chosen approach was better than the other options. Sometimes, particularly if the product development is unsuccessful, meeting notes can be used to help find where the process went wrong to avoid making the same or similar mistakes in the future.

Once the team is assembled, meetings are conducted on a regular basis to design and begin development of the product. One other responsibility of the product manager is that of producing the PDR from which the development of the proposed product will be guided.

Conduct design reviews

The design of a product is generally carefully planned and executed. It should reflect all of the features stated in the PDR. It should also meet as many as possible of the audience and customer needs and wants listed in the MRD. To ensure that the product meets the features, wants, and needs listed in the PDR and MRD, the product development team conducts design reviews.

A design review is a meeting in which all interested personnel look at each product feature and analyze how well the feature meets the needs or wants it was designed to meet. The design review discusses each of the various options that were considered for each product feature, and explains why the chosen option is the best one available under the current circumstances. Design reviews might also provide input about the product that was not previously available, but which must now be considered.

Design reviews usually include several or all of the product team members, and can also include members of the company's management staff. The primary purpose of a design review is to make certain that the product being designed is feasible, cost effective, and of sufficient quality.

Design reviews can be conducted on different aspects of the product resulting in several different design reviews. They can also be conducted on the product as a whole. Companies sometimes conduct partial design reviews as the product is being designed and developed, and then conduct a complete product design review once the design is established.

Design reviews can be conducted either before or after the first prototype is developed. If conducted before the first prototype is developed, the design review can provide input for the team members creating the prototype. If conducted after the first prototype is developed, the design review can be used to refine the prototype.

A product design review should be completed early enough in the product's design and development to prevent costly mistakes or misdirection in the design of the product. Design reviews can also be conducted later in the product's development to ensure that the product's design still meets the needs and wants of the audience and target customer.

Later design reviews become particularly important when the design and development of the product is lengthy. The likelihood that audience and customer needs and wants might change increases with the length of time it takes to design and develop the product. Design reviews conducted at the later stages of development, and which are compared to updated market research, can help to ensure that the product will still be marketable by the time it actually reaches the marketplace.

Prototype the product

As the product's design and development continues, there will come a time when the product is far enough along to allow the creation of a first product prototype—an example of the finished product. The prototype is often the first time the team sees the results of their efforts. It might also be the first time the team sees whether or not their efforts are likely to be successful.

In addition to letting the team see the potential results of their efforts, the prototype serves other functions. It can also be used for testing customer satisfaction, as well as for providing a starting point from which to begin the product's documentation.

The performance of the product in the hands of the consumer is a key indicator of the capability of the development, production, and marketing system. The successful manufacturing enterprise does not, however, simply wait to see emerging trends in market share and then respond. A constant assessment of the relative merits of competing products and the continuous incorporation of the appropriate responses into one's products are a mark of the successful firm. (Compton et al. 1992)

As W. Dale Compton indicates, not only must your product perform appropriately for the development team, it must also perform successfully for the audience and customer.

There are, of course, additional costs associated with demonstrating prototypes to prove the value and success of your product for your audience and target customer. However, those costs do not necessarily have to be exorbitant. A little creativity and patience can go a long way toward reducing or containing those costs.

For example, the McDonald's corporation began testing its breakfast menu by introducing it into franchises in rural areas. It proved to be so successful that it was not long before other franchisees decided to include the breakfast menu in their McDonald's as well. As of 1990, the breakfast menu accounted for 35 to 40 percent of McDonald's revenues (Griffith 1990).

McDonald's used patience in introducing their new breakfast product line. Zoom lenses for cameras were introduced with creativity. A prototype of the zoom lens was made for under $500, and it was then introduced and test marketed at a dinner party. "When the product is right, you don't have to be a great marketer" (Griffith 1990).

Prototype testing is often conducted to determine whether or not the product being designed and developed meets audience and customer defined needs. That is, testing is often performed in search of an appropriate response "within the context of the *needs and wants of the customer*" (Compton et al. 1992). The appropriate response often indicates whether or not the product can be easily used by the audience or customer. This type of testing is known as *usability testing*. While usability testing is considered to be one type of product testing, it is also a type of documentation testing. Usability testing will help to determine whether or not the product's interface is designed to suit the intended audience and customer.

The prototype created depends, of course, on the product. For example, a prototype for a software product such as a word processing program would look and respond the same way the finished product is intended to look and respond. Prototyping also points out areas where the product must be changed. Changing the product then results in additional prototypes. During the product's design and development, several prototypes might be created in order to determine the final and best design.

The major difference between early prototypes and later prototypes is often the use of created rather than actual product input. In the case of a word processing software program, created data instead of actual data would likely be used in the early prototypes. Later prototypes should use actual data to ensure that the data being used reflects the stresses and maximums the product will face once it is sold and in use.

In addition, only some of the features of the product might be available in early prototypes. A feature that allows one word processing file to be joined to another, for example, might be set up for the prototype to demonstrate the feature. By having only limited features completely functional in early prototypes, features can be demonstrated earlier than the entire product might otherwise be available.

Prototyping is part of the development of a product. It also has an important influence on the product's eventual design. The use of prototypes in the development of a product ultimately leads to a more-complete and better-designed product.

Development

The actual development of the product is a continuing process that begins once the product's design has been sufficiently established to allow for the product's creation. Often the first substantive indication of product development is the production of a product prototype. Development continues on a product from the first prototype to the first finished product. In between, there might be several prototypes.

A product's development is started once documents such as the MRD and PDR outline what the product is intended to do or be, and who the product is intended to be sold to. As mentioned previously, there are also documents that provide more detailed information about the interface the product will show to the audience and customer (user interface), or the way in which other product engineers can use the product to create their own products (engineering interface). These two documents are most commonly used by computer software manufacturers.

As product development continues, other documents are written to aid the team and other interested parties in understanding and detailing the full design of the product. These types of documents perform different though often similar functions throughout the product's development cycle. Three typical product development documents that might be created during the product's development are the

- Architecture document
- Design document
- Functional specification

As with the MRD and PDR, each of these documents can have different titles depending on the company producing them, the type of industry, and the type of product. You might find documents similar to these in your own company. Although these documents have different titles, you will be able to recognize each one by the type of information it contains.

The architecture document describes the overall design of the product. It includes information about each of the products major components and explains how these components fit together to form the final product. The primary purpose of the architecture document is to help designers and developers think through all of the individual features the product should contain in order to ensure that it addresses the needs and wants of the product's intended audience and target customer.

The architecture document has as its secondary purpose that of keeping all other teams and individuals who need to know about the product informed of its features. It helps those individuals not directly involved in the product's design to quickly come up to speed on its design and features.

As with other documents created during the planning, design, and development of a product, the architecture document should provide some basic items of information:

- A brief statement of the product's main objective
- A description of the product's intended audience and target customer
- A list with brief explanations of any assumptions being made related to this product, as well as any predefined product development criteria
- A description of each of the products major or key features
- Explanations and descriptions of how each of the product's key features relate to each other to create the final product
- A list of all other components, teams, products, and so on which this product depends for its successful completion

The design document provides some of the same basic information provided by the architecture document, such as a brief statement of the product's main objective and descriptions of each of the products key features along with explanations of how these key features relate to each other. However, the design document does not include information such as descriptions of the product's intended audience and target customer, or explanations of any assumptions being made related to this product, or any predefined product development criteria.

The design document's primary purpose is to provide details related to the product's design and information about how the development must proceed in order to complete the product. For example, if the product is a software application program, the design document should include details such as the structure of data to be used for this product, any programming interfaces that will be used, as well as any programming algorithms, parameters, and return values.

Regardless of the product type, this document should also include information such as what provisions will be made for testing the product, and how the product will be packaged for shipping. It should also include any other types of information that designers and developers might need to know.

The functional specification describes the features and use of the product once its design is complete. It draws much of its content from the design document, including descriptions of each of the products key features along with explanations of how these key features relate to each other.

While the design document concentrates on providing information about the product as the designers and developers intend to create it, the functional specification describes the product as it finally exists. The two documents might contain a great deal of the same information, but the functional specification is the final description of the product.

It is often from the design document and functional specification that those who create the end-user documentation obtain their information. The primary writer will also gather much of the information needed to write the documentation from the design review meetings as well. Taking these two approaches to gathering product knowledge and information from which to create product documentation helps to ensure that the documentation is reasonably accurate. Even so, problems do find their way into product documentation. The only way to make sure that the documentation is technically correct and usable is for a documentation tester to be involved as a member in both the product and documentation development teams, and for the tester to conduct one or more tests on the final product documentation.

The development of the first finished product and the documentation that accompanies it leads to the next phase of product creation—product testing. Product testing and documentation testing are often conducted simultaneously.

Product testing

While some might consider design reviews to be part of the product testing cycle, they are really a separate process conducted while the basic design of the product is still in flux. Product design reviews are intended to ensure that, during each step of the design process, the product successfully meets the features, needs, and wants outlined in the MRD, the PDR, and other internal product documents, and that the product does so in the most efficient and effective way possible.

To some extent, the purpose of product testing is very similar to that of conducting design reviews. Product testing must also ensure that the product meets the features, needs, and wants outlined in the internal product documents. The main difference is that product testing is conducted to make certain that the final product has not strayed from its goal from the standpoint of the audience and customer, rather than from the standpoint of the designers and developers. This is also one of the reasons for conducting documentation tests.

To ensure the validity and usefulness of the product from the audience and customer point of view, product testing is generally conducted by individuals who, while they might have participated on the development team, were not primarily responsible for the design and development of the product. This is not to say that product testers do not have input in the product's design and development; it is to say that designing and creating the product is not their main or primary responsibility. In addition to participation on the design and development team, product testers have four other responsibilities:

- Developing test requirements
- Conducting product tests
- Reporting test results
- Conducting and passing on final product test

Developing test requirements

Depending on the product, establishing test requirements can be the easiest part of the testing process, or the most difficult. Most often, creating test requirements is the most important part of product testing. Unless the test requirements are correctly identified and established, the actual testing will be of little or no use.

There are different methods for determining test requirements, sometimes referred to as *benchmarks*. According to W. Dale Compton,

> *In some industries it is common to emphasize a few operating characteristics, while in others the focus will be on a wide range of parameters. In some cases the parameters that describe system performance will be highlighted. In others, it may be more common to examine the performance of subelements of the system. Some metrics may be reasonably easy to obtain. Others may need to be estimated based on limited available data. However it is accomplished, benchmarking is a critical foundation for the successful operation of a manufacturing enterprise.* (Heim and Compton 1992)

Conducting product tests

Once the testing requirements are established, and the first design of the product is complete, product testing can begin. The testing follows the guidelines and requirements set out in the testing requirements document. That document should list each type of test that will be conducted, what special equipment will be used to conduct the test (if any), under what circumstances the product is being tested, what results are required in order for the product to pass the test, and any other information that relates to the specific product being tested.

Product and documentation tests are often conducted simultaneously. In addition, both require that a testing plan or document of some type be created before the test is conducted in order to establish the testing parameters and point out those factors that indicate a successful test. Product and documentation tests have one other major item in common, they are both conducted based on the actual product created.

Testing is a reasonable method for ensuring that the product, and the documentation, meet quality standards established by the company. While product testing is important to verify that the product functions as designed and developed, documentation testing is important to verify that both the product and its documentation are sufficiently successful and important to the customer so as to generate demand for the product.

Don Debelak uses a comparison example to show how important the ultimate benefits provided by the product are to the customer. Although not originally intended, his example also shows how important testing is to the quality of the product.

> *Mosquitos are a constant hot-weather problem. Just outside my front door in summer, I often have anywhere from 20 to 100 mosquitoes hovering, waiting for their next meal. When a company introduced a small electronic device that repelled mosquitoes, I bought one for $5.95. The product's benefit was very appealing to me.*
>
> *Another inventor had devised an innovative spice rack that was designed to be pulled in and out of a kitchen cabinet. The product was an improvement, but because its benefit wasn't important to most people, the product failed.* (Debelak 1992)

To Debelak, the electronic mosquito repellent was a successful product. Product testers would have established and worked the product against a series of quality control tests to ensure that it functioned as designed—zapping mosquitoes—before the product was released to the market. The product testers would have proven the quality of the product as built. However, Debelak makes another statement that indicates what the product testers would not have found. Debelak also states that

> *When consumers look at a product's benefit, what matters is how much better that benefit is than the benefits of competing products. The mosquito repeller's competitors were bug sprays. By comparison, the repeller didn't smell, lasted forever, and was environmentally safe. The spice rack's 10 to 20 percent improvement wasn't enough to attract customers.* (Debelak 1992)

If the original designers had been more concerned about creating a new bug spray that was odorless and environmentally safe instead of coming up with the idea for an electronic device, they would have produced a product whose benefits were similar to their competitors. It is likely that their final product would have been no more than 10 to 20 percent better than the competition, and thus not sufficient to attract customers.

Product testers would have tested this product for functionality, and as long as it met its predefined criteria (kills mosquitoes, does not smell, and does not harm the environment) the product would have passed its product testing. Those who designed and built the product, as well as those who tested it, might never have realized that while they had built a better aerosol repellent, that was not really what the market wanted. On the other hand, it is a documentation tester's responsibility to clearly understand what the market wants and needs, and to test the documentation to, among other things, make certain that it meets those wants and fills those needs.

One way the documentation tester accomplishes this is by creating documentation test plans that look for and test the success the product has at meeting customer needs and wants. Because the documentation tester is not as close to the product's design and development as is much of the rest of the team, the documentation tester can more easily keep himself and his tests focused on the customer's view of the product.

Reporting test results

Regardless of whether the tests conducted were done so to test the product or its documentation, the results of the test must be clearly and effectively reported. In some product development environments, the test plan is reviewed by development team members and approved before any testing ever begins. In this environment, all team members know what is to be tested, how it is to be tested, and what the final results should show. The report that must be created in this case can concentrate on what portions of the test the product did not pass, as all involved parties already know what was tested and how it was tested. When product testers are less formal at reporting and obtaining approval of their approach to product testing, the final report must also explain what was tested and how the test was conducted, in addition to reporting the outcome.

The testing report does not necessarily have to be in written form. A verbal accounting of the tests conducted and the results of those tests can be presented at a team meeting. The advantages of reporting the test results in a more formal written document, however, often make the added work worth the effort.

One good reason for providing test results in a formal written report is that it provides a product testing history. This information can be used to redesign and improve the product, or possibly even to stir ideas for a different but similar product. Without a written report, much of this information might be forgotten or lost when the product tester moves on to another company.

Another reason for providing written test results is that it provides an outline of what you do not have to test the next time this product is revised. Improved products with only minor product revisions do not have to be completely retested if the results from the original tests are available for review.

The software industry provides the best example here. When an application program is first created, every aspect of that program is thoroughly tested to ensure that it performs as intended. Even the screens that the user sees are tested to help provide the best possible user interface. When the application program is later modified, only certain portions of the program are changed. Those portions that are not changed do not have to be retested if the results of their original tests are available for review. There is one caveat here, however. In software programming, one minor change to a small section of the program code can result in major malfunctions in other areas of the code. To be on the safe side, many software product developers retest all of the code to ensure it continues to function as intended. Even in this circumstance, however, it is often possible to test only the major functionality of the unchanged code as a test case to verify that a particular portion of the unmodified code did not malfunction due to other code changes. This approach can substantially reduce the hours required to test a changed product.

Conducting and passing on final product test

Even though a product might be thoroughly tested during its development, some products are given a final test. This type of a test is generally quite superficial, and is intended to make certain that slight corrections and modifications made as indicated by previous product tests have not "broken" the product as a whole. This type of test is common to software development. But even in other types of products, this type of test provides one last chance to find and fix anything that might result in product failure.

Once the product successfully passes this final test, the product tester signs off on the final testing document, indicating the product is ready for the next phase—product release.

Product release

Product release is actually a two-part phase within the creation and development of a product. The first part includes the manufacturing of the product. The second part includes the distribution, shipping, and selling of the product, also known more generally as product release. Regardless of the type of product, all nonservice products go through this two-part manufacturing and release phase.

Product manufacturing

While the design and development team conceives, creates, tests, and produces the basic product so as to ensure its successful acceptance in the marketplace, its efforts can be impeded or even completely eliminated during the manufacturing phase of the product. Regardless of how well designed and tested a product might be, if problems occur during manufacturing, the success of that product can be in great danger.

The primary concern here is the loss or reduction of product quality during manufacturing. Several factors can result in product quality reductions, or if implemented properly, product quality maintenance (Compton et al. 1992):

- Changes in the efficiency of labor
- Changes in the manufacturing process
- Modifications to the product's design
- Standardization of the product
- Changes in volume production
- Substitution of materials in production

Manufacturing can be a tedious process, as well as a boring job. Counteracting the tediousness and boredom can change labor efficiency. One way to counteract these potential quality robbers is by providing training and incentives. Training on the product and its importance to the marketplace helps to improve employee interest in the product and, thus, their attention to detail. Incentives for quantity and quality also help to increase employee efficiency.

Changes in the manufacturing process can either improve or reduce the quality of the product. If not made with sufficient thought as to the effect that changes will have on the final product, changes can greatly reduce the quality of the finished product.

Modifications to the product's design that must be made by manufacturing in order to accommodate their ability to produce the product can also have negative or positive effects. If the modifications are made after the product's original design and development has been tested and approved, the chances are that product design changes will do more harm than good. If, on the other hand, product design changes are made during the original design and development of the product, the design changes might have no negative effect on the product, and could in fact improve the product for its audience and target customer.

Standardizing the product so as to reduce the variety of tasks that must be performed by labor in the manufacturing of this product can improve the quality of the product. At the very least, it counteracts the problems that might be associated with not standardizing the product, the most obvious of which is an increase in the variety and number of tasks that must be performed in order to produce the product. Any time you introduce additional tasks or steps in the manufacturing of a product, you also introduce more points in the process for something to go wrong. Even slight problems can reduce the quality of a product.

Changes in volume production can also have a negative effect. For example, large increases in production that were not originally anticipated can result in problems such as machine variances that are no longer within required limits or a lack of sufficient raw materials of the specified quality. On the other hand, volume increases that are planned and designed for can reduce the overall cost of manufacturing without reducing the quality.

Materials substitution can also be a two-edged quality sword. Substituting raw materials can maintain or increase a product's quality, providing enough time and research are dedicated to finding lower-cost materials with equivalent or greater quality. Materials substitution can be a serious problem, however, if you are being forced to substitute a lesser-quality material to keep up with product demand or to maintain production costs.

Besides potentially impeding the successful release of a product, manufacturing can also contribute substantially to its success. One way in which manufacturing can contribute to a product's success is by participating in product design and development meetings. Marketing people can provide the details about a product's potential audience and target customer. Engineering (design and development) people can provide the technical expertise necessary to turn customer problems into potential solutions. Manufacturing people can provide the creativity of manufacturing alternatives that contribute to the most effective production of a product.

As engineering and marketing go about establishing their product definition, somewhere there exists a role for manufacturing. Depending on the products and the expected cus-

tomer benefits, manufacturing can be brought in at a variety
of points in the process—usually the sooner the better. This is
not to suggest that manufacturing will play as critical a role
as marketing and engineering, but there are quite a number
of products where some manufacturing creativity can, and
has, dramatically improved the customer's acceptance of a
product. (Edmondson 1992)

Besides death and taxes, there is one other thing on which you
can depend—things change. In manufacturing, as in other aspects of
business, costs change, and for the most part they tend to increase. If
cost increases are planned for when a product is designed and developed, the probability that a product can continue to be manufactured at a quality and a cost that make it profitable is improved.
Because things change, changes that have the potential to negatively
impact a product's quality also have the potential to positively impact
a product's quality, or at the very least, to maintain it at a suitable
level.

For example, changing the efficiency of the manufacturing labor
force can also reduce the cost of labor, or at least maintain it at its current level for a greater period of time. In addition, changes in the
manufacturing process can also contribute to reduced or stabilized
manufacturing costs, as can making modifications to the product's design. Standardizing the product, including its design, can also improve its quality and reduce the costs associated with producing it. In
addition, both changes in production volume that are planned and
prepared for, particularly increases, and the flexibility to substitute
other equally suited materials in the production process can contribute to quality maintenance or improvement.

Product release

Products are often designed, developed, tested, and manufactured
with a single target goal in mind—getting the product into the hands
of that first customer by a predetermined date. In the software industry, that date is often referred to as the FCS (first customer ship) date.

The FCS date is the culmination of months, sometimes years,
worth of research, planning, designing, developing, and other related
tasks. Although usually officially deemed complete at FCS, there often are some tasks that still must be performed. The most important
is that of the postrelease review.

The delivery of any new product carries with it the responsibility
of determining whether or not the process required to deliver that
product was as efficient and effective as it could have been. In the

case of a successful product, it will most often be determined that much of the process went smoothly.

Whether the product is a success or a failure in the marketplace, it is important to review the entire process. Process review shows the company what it did right so that it can repeat those successes, and what it did wrong so that it does not repeat its mistakes. Process review also helps the company determine one other important item—the product's opportunity cost.

Opportunity cost "extends to include sacrifices that are made by foregoing benefits or returns. Opportunity cost takes into consideration the fact that choosing one of several alternatives precludes the receiving of the benefits of the rejected alternatives" (Stevens and Sherwood 1982). Determining the opportunity cost associated with a company's final product completes the picture of the product's total cost. It helps a company see whether or not it made the right decision. In addition, it might also help the company develop another suitable product, the one that was not created this time around.

One other task that should be performed at FCS is watching the success or failure of the product. Regardless of whether the product succeeds or fails, correctly assessing the reasons for its final results in the marketplace puts you on the right path toward improving or replacing that product, whichever is deemed most appropriate. In other words, it starts you back at stage one of product creation—product research and planning.

An overview of the documentation development process

Documentation development begins almost as early as product development, even though the writing of the actual documentation does not begin until much later. The documentation development process, like the product development process, begins with the initial research stage.

Initial research

The initial research stage for product development includes creation of two documents—the MRD and the PDR. Later in the development stage, an architecture document, design document, functional specification, and other related documents are also produced.

The initial research stage for documentation development includes the review and study of all documents created for the product. In addition, the initial research stage for documentation development includes

- Review and study of documentation from product competitors
- Review and study of previous versions of the product documentation

The documentation writer, editor, and tester should be involved in this and subsequent stages of product documentation. Writer, editor, and tester involvement also requires and assumes their involvement in the design and development of the product, as previously discussed.

The documentation development team consists of three or more persons responsible for performing the specific functions related to creating product documentation. These functions include writing, editing, and testing the product documentation. This team is referred to in this document as the WET team (Fig. 2-3).

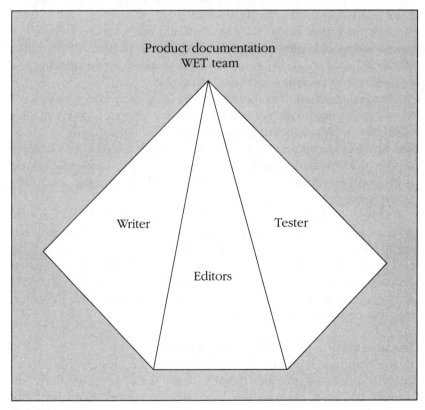

2-3 *The WET team*

It is the WET team's responsibility to develop the product's documentation and ensure that it meets the needs and wants of its audience and target customer. In addition, the WET team is also responsible for ensuring that the product documentation is complete, accurate, and of the highest possible quality given the development time frame and dependencies.

The WET team begins to accomplish their task of creating complete, accurate, and high-quality product documentation by reviewing all existing relevant product documentation. This documentation includes company internal documents created for this product, competitor's product documentation, and any company documentation that exists for a previous version of this or a closely related product.

Internal product documentation

The most common types of internal product documentation available for a company product were described in detail earlier in this chapter; they will not be described again here. There are two items of information about internal product documentation, however, that are particularly relevant to the WET team.

First, members of the WET team should read and study the company's internal documentation. It provides the product details team members need to know in order to be successful at producing the end-user (customer) product documentation.

Second, internal documentation rarely, if ever, undergoes documentation testing, with one exception. Most internal documentation is subject to an engineering review type of documentation test. Unless the documentation tester is, however, both the product engineer and the documentation tester, this type of internal documentation test is generally conducted by the product's engineers, not the documentation tester.

This documentation provides the WET team with information such as a description of the product, details about the market needs that it must meet, and the time frame within which the product must be produced in order to take advantage of the market's desire for the product. It also describes each of the product's features, and gives the writer a good base from which to begin planning the documentation that will be needed for the product.

Documentation of product competitors

Reading and studying the documentation of the proposed product's likely competitors provides the WET team with two particularly important pieces of information. First is information about how the competition chose to pursue the development of product documentation. This knowledge helps the WET team see one or more approaches to product documentation. A little additional market research (referencing product reviews written about the competitors' products and their related documentation) will then tell the writer whether or not any of the competitor's approaches were particularly effective, and if so, which ones.

Second, reading and studying the competition's documentation also lets the WET team compare product features. By understanding the competition's product features, anything that seems to be necessary but missing from the design of this product can be brought to the development team's attention by one or more WET team members.

Previous versions of product documentation

The WET team members should also read and study any previous versions of documentation available for this product. Besides the product knowledge this approach provides, it can also point out areas where the product documentation was particularly effective, or where it did not produce the intended results.

Studying this documentation and comparing it against the information gathered in other areas of the company—service and support and complaints departments in particular—the writer has additional useful information to help him design and produce the best and most useful product documentation.

Documentation plan

Once the initial research has been conducted and all relative information gathered, a document outlining the product documentation to be produced should then be developed. This type of document is referred to as a *documentation plan.*

The purpose of the documentation plan is to guide the writer in the development and writing of the product documentation. Separate documentation plans can be created for each book that is to be written. Sometimes, however, only one documentation plan is created. It includes all books or documents to be written for the product. Regardless of the approach taken—one book per documentation plan, or multiple books included in one documentation plan—the basic information contained in the documentation plan is the same.

Documentation plans commonly provide several types of information (Woolever and Loeb 1994):

- A list of books being created or the title of the document being discussed in the documentation plan
- A table of contents and an overview of the documentation to be created

- A definition of the audience and target customer
- Outlines of each document
- Information about related documents
- Production details
- Completion needs and requirements
- Documentation schedule

Document and book list
This section of the documentation plan includes either the proposed title of the document to be created or a list of titles if there is to be more than one document. In addition, document revision numbers should be included, along with revision (version) history information if appropriate.

Table of contents and documentation overview
A table of contents for each document is included in this section. The table of contents should be as thorough as possible, including appendices, chapter titles, subheadings, etc.

Audience and target customer definition
To justify why you are preparing certain types of documents, and the approach you are taking, you should include a description of the documents audience and target customer. This information is available in both the MRD and PDR.

Document outline
This section should tell the documentation plan's readers what will be included in each chapter of each document. A paragraph description, a multiple-level outline, or any other approach is acceptable, as long as reader's understand the level of information that is being created.

Related documents
If the document described in the documentation plan is one of a larger set, or part of internal documentation, this section describes how the document relates to other documents. In addition, this section contains information explaining how the proposed documentation relates to competitor's product documentation. Take the opportunity to describe how this documentation will be better than any competitor's product documentation. You can also describe how this documentation will improve on previous versions of your company's existing documentation.

Production details

Information related to any production the writer is responsible for should be included in this section. If delivery of camera-ready copy is the only requirement, this section should indicate that. If the writer is not responsible for production, this section should include the total number of pages anticipated or required for each document, and might also include the typeface or font to be used, how graphics will be provided, and how camera-ready copy will be provided.

If the writer is responsible for production of any document described in the documentation plan, information such as the following should be included:

- Number of pages
- Typeface or font
- Graphics
- Binding
- Colors
- Paper stock
- Production type

Completion needs and requirements

This section should contain a list and, if necessary, descriptions of any resources that will be needed in order to complete the documentation. For example, if a specific kind of word processing software or personal computer is required, those should be included in this section. If you need to have copies of all internal product documentation, this information should be included along with, if known, a list of those documents. If you need to have the product in hand, this should be included here as well. Anything that the writer might need in order to successfully complete the product documentation should be included in this section.

One other item that should be included in this section, where applicable and if available, is a cost estimate for the creation of the documentation. This is particularly important when the documentation plan will also be used for budgeting purposes, or to convince management that taking one direction with the documentation is more beneficial than taking another direction (such as on-line documentation instead of hard copy documentation).

Documentation schedule

This section can be presented in any format, but must specify the milestone dates and complete timeline for each document that will be produced. An overall timeline for all product documentation is a good starting point if detailed milestones and dates cannot be included for each separate piece of documentation.

Document outlines

Part of the documentation plan includes the development of outlines for each document to be created. Whenever possible, those outlines should be very detailed. However, at the early stages of documentation development, the documentation plan might only include single- or dual-level outlines. Once the documentation development process proceeds a little further, however, more detailed document outlines are required.

The documentation tester also has a place in the preparation of document outlines. When a writer and documentation tester work closely together from the beginning of the documentation development stage, the documentation tester can contribute by reviewing the outline and helping the writer find any areas where information is missing. In addition, the documentation tester can conduct a simple read-through documentation test at this stage, also to help find any areas where engineering changes have been made to the product but have not yet been reflected in the document outline.

Creating the initial outline draft is usually possible because of the information gathered from reading the initial technical documentation. To provide a more complete draft, however, generally involves preparing a list of additional questions and conducting interviews.

Document drafts

Once the detailed outline is complete, the next step is creating the first draft. The first draft should contain as much information and be as well written as possible.

> *Just as an engineer must understand material properties; a systems programmer, console modes; a field service rep, diagnostic functions; and a marketeer, the market, so must a competent writer understand rhetorical devices in order to achieve the most elegant and proper solutions. Understanding grammar, its usage and structure, is basic to all good writing.* (Bell and Evans 1989)

In addition, the first and all subsequent drafts should also be as technically correct as possible. Despite the writer's best efforts, however, documentation often still contains errors. To help ensure the technical accuracy of written documentation, documents should be given to product developers for review. Most document drafts are distributed to all members of the development and documentation teams. They can also be distributed to other interested parties, by special request.

After product developers review the documents, their changes and suggestions are incorporated as appropriate, and another draft is created. Several drafts are often required, with each draft being reviewed for accuracy by the product's designers and developers.

Document edits

Several edits are often performed on document drafts as well. Document edits are also conducted at different levels. Two common levels of document edits are referred to as copyedits and substantive edits, although other titles often apply. The copyedit and substantive types of edits are also sometimes referred to as thorough editing, cursory editing, production editing, and minimal editing (Boston 1986).

A *copyedit* is the final type of edit performed on documents. Its purpose is to correct spelling, grammar, and punctuation errors, as well as to ensure that the document is consistent. A copyedit also includes verification of the document's completeness, accuracy, and the format of tables, bibliographies, footnotes, and so on. A copyeditor does not rewrite or reorganize the document being edited. By the time a copyedit is performed, all parties should be comfortable with the organization and style of the document.

A *substantive edit* includes not only copyediting, but also includes rewriting, reorganizing, and a complete review of all aspects of the document. Editors write transitions and chapter or section summaries if needed. Editors also edit the document to eliminate wordiness, ensure the proper tone has been established and carried consistently through the document, follow the document's flow to verify accuracy and logic, and check to ensure the document is correctly written for its intended audience.

It might take several revisions and drafts to create a document ready for final editing. Once final editing is complete and changes are input, the document is ready for documentation testing.

With all of the drafts, edits, and reviews that a document goes through, it seems as though the document should be so complete and technically accurate that no documentation test would be required. Unfortunately, this is not usually the case.

Developers often cannot give the document the thoroughness of review that is necessary in order to find and correct technical errors or inaccuracies. Hanna Bandes offers the following explanation for why product developers often do not or cannot provide the kind of detailed technical review that is required to ensure a document's complete technical accuracy:

> *Often reviewers just read through a document, changing grammar, spelling, and punctuation. Rarely do they look at the document's organization. Although they often note inconsistencies or inaccuracies in the material before them, only rarely do they notice that material is missing. And a manual that is missing important information fails in the areas of completeness and usefulness.* (Bandes 1987)

Even editors performing very detailed edits cannot always determine when information is missing or incorrect, particularly in fields where the product is very specialized or technical. One reason for this is that editors are generally hired for their editing skills, and not for their product knowledge. They are more concerned with whether or not sentence structure, punctuation, logic, and other related items are accurate. In defense of editors, that is what they are hired for. They are doing their job when they find and correct errors or problems of that nature. They are not hired with the expectation of complete product knowledge.

Editors are also rarely, if ever, included in product development. They do not attend development meetings, work with the product prototypes, or otherwise become very involved in the workings of the product.

Good document reviews are an important part of producing good documentation, no matter who is doing the reviewing. But they can be very hard to get. Bell and Evans (1989) state that, "Good reviews are not 'hey there, great job,' but are the result of careful reading, meticulous editing, and timely return of documents." Rarely is it possible for all development and documentation team members to meet this definition of a good review. The problem often comes down to limited time, lack of resources, and the perspective that the product is very important, but the documentation is only a necessary evil. Bell and Evans provide additional support for this concept in the following paragraph:

> *When the draft appears as clean as it went out (except for a typo or two that the reviewer has spotted), the reason is most likely not a perfect document, but a reviewer whose manager has not made it clear that the documentation is important by providing the perqs to prove it.* (Bell and Evans 1989)

All of these factors show the importance of documentation testing. A company that is truly dedicated to product quality will also be dedicated to the quality of its accompanying documentation. It will be a

company interested in developing and establishing a TQM (total quality management) system, as described by quality experts Frank Mahoney and Carl Thor in their book *The TQM Trilogy* (Mahoney and
Thor 1994). Such a company will often seek to obtain one or more
quality certifications or awards, such as ISO 9000 certification, the
Deming Prize, or the Baldridge Award. Such a company will recognize
the importance of producing the right kind and right amount of documentation. They will also make every effort to ensure the best possible
quality of their documentation by integrating documentation testing in
their product and documentation development processes.

Document tests

As chapter 1 explained, a documentation test consists of six major
steps or phases performed by a documentation tester:
* Participating in product and document development
* Planning and preparing for the documentation test
* Conducting the test
* Reporting the results
* Following up after the test
* Following up after product release

During these steps, the documentation tester's primary mission
should be that of contributing to the production of the highest-quality document without interfering with or lengthening the production
schedule. The documentation tester's early and continued participation in both product and documentation development helps to ensure the tester fulfills that primary mission.

Of course, the mission is not fulfilled until the documentation
tester has completed the most appropriate test possible within the
time limits and other restrictions placed on both the documentation
and the tester. In order for the tester to do that, the tester must choose
the most appropriate test and conduct the test as scheduled.

The tester has several types of tests from which to choose:
* Read-through
* Engineering review
* Usability
* Basic functionality
* Integration

Whatever test type is chosen for the documentation, its timely and full
completion is an important part of the documentation process, and
should not be overlooked. Some testing of the product documentation should be conducted, even if there is not sufficient time or resources with which to conduct the best or most-effective test.

Writer review

Once the documentation test is complete and all edit, review, and documentation test changes have been input, the writer should conduct a final review of the documentation. If time permits, the writer might choose to read the document from beginning to end. If time does not permit, the writer should attempt to at least spot check the document, particularly after the master copy is printed for production.

One other item the writer should be looking for in particular is the accuracy and usefulness of the document's index. Many organizations assign the task of indexing the document to someone other than the writer. The documentation tester rarely gets to see the index, as it is usually not produced until the document is complete, page numbers are fixed, all changes are input, and so on. Therefore, the only person who gets the opportunity to verify the index might be the writer. As one final check, the writer should attempt to look up some of the index references. If the writer finds that even one reference is incorrect, the entire index should be checked.

Once the writer makes a final review of the document, it is then ready for production.

Production

The production process can be as simple as making a few photocopies of the final document, or as elaborate as producing thousands of copies of a multipage, bound, color document. It can even include producing a CD-ROM.

Some companies produce the document themselves, while others contract the work to another company. Regardless of the approach, the minimum that must be done is to prepare the final copy of the document for production and then oversee its final production.

Even after the document is produced and ready for shipment, the documentation development process is not quite complete. Just as the tester conducts or participates in a postrelease product and document review, so too must the documentation team.

Postrelease review

The purpose of the postrelease review is to determine what went right and what went wrong in the development and production of the product documentation. It should also be a time when several other things are accomplished.

If postrelease information about the documentation is available, this information should be reviewed. Some types of information that might be available include a list of service and support calls taken as a result of one problem or another with the product's documentation. Also, if the company has a complaint department, any complaints that were filed should be reviewed. In addition, if registration forms or customer satisfaction forms were distributed with the product, a summary of the results of those forms should be prepared and presented at the postrelease review.

Reviewing what did not go well during the document production cycle helps all team members improve the process next time around. This also means that besides discussing what went wrong, constructive suggestions for fixing the problems should also be solicited and recorded. In addition, it means that what went particularly right should be noted, and if appropriate, integrated into the company's standards and practices for the future.

The postrelease review is not intended solely to find what went wrong during the last release. It is also intended to point out what went right, and to reward the team for their success. To this end, postrelease reviews are often conducted off-site (a location other than the employer's place of business). Rewarding a team for a job well done is a small price to pay for the efforts made by each individual and the collective team on behalf of the company and its product. Appreciation for a job well done is never wasted.

How the documentation tester contributes to development

A documentation tester might become involved in the development of a product even while marketing is still attempting to define that product and the needs or wants it is to meet. Documentation testers often participate in information gathering conducted at the postrelease of a product. The information gathered in this manner can then be used as input for the definition of the next version of the same product or of a new product.

Documentation testers might help gather information from customers or potential customers by participating in many activities including

- Contacting potential or existing customers using questionnaires
- Participating in roundtable discussions with existing or potential customers
- Conducting telephone or electronic mail surveys
- Observing usability tests on products

Besides gathering information from customers, documentation testers might also speak with internal personnel and gather internal information including

- Service and support statistics
- Customer complaint information
- Product testing reports
- Other related documents

If a documentation tester does not participate in these activities, the documentation tester's initial participation in the product's design and development begins when presented with the initial internal documents. It is the documentation tester's responsibility to read and understand these documents. A documentation tester might even produce some of the initial documents by participating in document design and review including the

- Market requirements document (MRD)
- Product design requirements (PDR)
- Architecture document
- User interface design
- Engineering interface

As noted earlier, because a documentation tester does not become heavily involved in a product's development until its viability in the marketplace has been shown, the additional participation of a documentation tester on the product development team helps to ensure the buyer-oriented development of the product. Unless a product is developed with the buyer in mind, the product is not likely to be as successful as it otherwise should be.

One reason the documentation tester is important for ensuring the buyer's point of view is that the documentation tester generally has a broader perspective of the market, the product, and how it will be used than does any other member of the product development or documentation development teams. This is particularly true when it comes to the interaction between the product development team and the documentation development team. In some instances, the documentation tester attends product development meetings as a representative for the documentation team, as well as being a representative for documentation testing. In this situation in particular, the documentation tester sees the product from both the documentation and product points of view, and might be the only team member who does.

The documentation tester participates in the product's development at all stages including

- Assembling the team
- Conducting development meetings

- Conducting design reviews
- Prototyping the product
- Preparing the product for testing

As the team is assembled, the documentation tester needs to remind the product development manager that his services on the team are useful. When development meetings are conducted, the documentation tester should be present at all meetings. The documentation tester represents the needs of the writer, the reader, and the tester on the development team and at development team meetings.

When design reviews are conducted, the documentation tester should be on the list of those participating in design reviews. Not only will the tester learn more about the product, but the tester's input can be useful here as well.

As prototypes are created, the documentation tester can also provide input regarding the interface and functionality of the product. In addition, by acting as a guinea pig, the documentation tester can help the product's prototypers prepare for the "big" prototype presentation to management. A documentation tester who asks questions that a nondesigner might ask helps the developer see aspects of the product that might not have otherwise been obvious to them. The prototyper can then enhance or correct, as needed, to ensure that the presentation to management is as smooth and effective as possible.

In the process of preparing a product for testing, a product tester can compare notes with the documentation tester. Each can see what aspects of the product will be tested. This helps the product tester eliminate any duplication of testing that the documentation tester will be performing. At the very least, it makes it possible for the product tester to design tests without the aid of or need for documentation in mind.

During the development of the documentation, the tester participates in all phases. As with the development of the product, the tester should be counted as a member of the documentation team. The documentation tester can help in all of phases of document production including

- Defining the audience and target customer
- Outlining each piece of documentation
- Providing information about related documents
- Helping to gather information about the product and verifying production details
- Ensuring the documentation fulfills all needs and requirements as identified in the initial documents
- Contributing to the development of the documentation schedule

The documentation tester's primary goal is, of course, to ensure that product documentation is of the highest quality and is as free of errors as possible. The different types of tests available to a documentation tester give the tester a variety of approaches to use. And while the tester can help in many other ways, the documentation tester's primary goal is the successful conducting of the required documentation tests within the time and other constraints placed on him.

Summary

The development of any product begins with an idea or a concept. From that idea grows a development team, a documentation team, and ultimately, a finished product with accompanying documentation. In between lies a product development process that is directed toward producing a successful product, within a predefined time limit, at a cost and quality that will provide its company with a sufficient return for the time and money invested.

The product development process includes conceiving, designing, prototyping, developing, testing, building, packaging, and shipping the final product. While product documentation also has a process of its own—conceiving, outlining, writing, editing, and testing—it really is a part of the overall product development cycle, into which the documentation tester also fits.

As a company employee, a team member of both the product and document development teams, and an independent documentation tester, the tester can successfully contribute to almost all phases of product and document design and development. In addition, the documentation tester is the final checkpoint for ensuring the quality of the product documentation. Regardless of the other contributions the documentation tester and other team members make to the design, development, and production of the product and its accompanying documentation, the final test of quality for a product's documentation is not complete until the documentation tester finishes document testing and the comments, suggestions, and corrections the tester made are incorporated into the documentation.

References

Bandes, Hanna. "Designing and Controlling Documentation Quality—Part 2." *Technical Communication* (Second Quarter 1987): 69–71.

Bell, Paula, and Charlotte Evans. 1989. *Mastering Documentation with Document Masters for System Development, Control, and Delivery.* New York: John Wiley & Sons.

Boston, Bruce O., ed. 1986. *Stet! Tricks of the Trade for Writers and Editors*. Alexandria, Virginia: Editorial Experts, Inc.

Compton, W. Dale, Michelle D. Dunlap, and Joseph A. Heim. 1992. "Learning Curves." In *Manufacturing Systems: Foundations of World Class Practice*. Washington, D.C.: National Academy Press.

Debelak, Don. 1992. *How to Bring a Product to Market for Less than $5000*. New York: John Wiley & Sons.

Edmondson, Harold E. 1992. "Customer Satisfaction." In *Manufacturing Systems: Foundations of World Class Practice*. Washington, D.C.: National Academy Press.

Griffith, Joe. 1990. *Speaker's Library of Business Stories, Anecdotes and Humor*. Englewood Cliffs, New Jersey: Prentice Hall.

Heim, Joseph A., and W. Dale Compton, eds. 1992. *Manufacturing Systems: Foundations of World Class Practice*. Washington, D.C.: National Academy Press.

Mahoney, Francis X., and Carl G. Thor. 1994. *The TQM Trilogy: Using ISO 9000, the Deming Prize, & the Baldridge Award to Establish a System for Total Quality Management*. New York: AMACOM.

Stevens, Robert E., and Philip K. Sherwood. 1982. *How to Prepare a Feasibility Study*. Englewood Cliffs, New Jersey: Prentice Hall.

Wilson, Richard C. 1992. "Jazz: a Metaphor for High Performance Teams." In: *Manufacturing Systems: Foundations of World Class Practice*. Washington, D.C.: National Academy Press.

Woolever, Kristin R., and Helen M. Loeb. 1994. *Writing for the Computer Industry*. Englewood Cliffs, New Jersey: Prentice Hall.

3

Preparing for a documentation test

Documentation testing is part of both the product and documentation development cycles. Its function and goal is to contribute to the quality of both the product and the product documentation, although its primary emphasis is on product documentation.

Testing is not a new science. The word *test* itself comes from the Latin word *testum* for an earthen pot or vessel. Pots of this nature were used to assay metals to "determine the presence or measure the weight of various elements." References to product-associated testing go back to 1950, indicating that it "was a routine activity associated with engineering and manufacturing processes, and it was quite natural to see it take shape as part of the software development process" (Hetzel 1988).

It is also quite natural to see testing take shape as part of the documentation development process. In software development, testing is a process of running a program or a system with the purpose of finding any errors it contains. A similar purpose holds true for documentation testing as well—using the documentation as designed with the purpose of finding any errors, omissions, or other potential problems.

Bill Hetzel provides an excellent summary related to software testing which, with a few relevant word changes, applies equally well to documentation testing. For documentation testing, Hetzel's revised summary would read, with this author's changes in brackets, as follows:

1 [Documentation] testing is any activity whose aim is the measurement and evaluation of [documentation] attributes or capabilities.

2 [Documentation] testing is an information-gathering activity.

3 [Documentation] testing is the measurement of [document] quality.

4 Reviews, inspections, and walk-throughs are important "tests" of the early phases of [document] development.

5 The central concerns for any testing method are what to test, when to stop, and who does the work.

6 Most [documentation] testing done today is ad hoc and poorly systematized.

7 The purposes of [documentation] testing are to
   • Gain confidence that [documents] may be used with acceptable risk.
   • Provide information that prevents errors from being made.
   • Provide information that helps detect errors earlier than they might have been found.
   • Discover errors and [document] deficiencies.
   • Discover what system [and documentation] capabilities are.
   • Provide information on the quality of the [documentation associated with] software [and other] products. (Hetzel 1988)

With this rather concise (and somewhat revised) summary of documentation testing in mind, the documentation tester can prepare for, choose, and conduct the most appropriate product documentation tests. This chapter shows how to prepare for testing product documentation, beginning with why it is important to become an active member of the development teams, and ending with suggestions for final preparations that must be made in order to ensure a successful documentation test.

After reading this chapter you will
   • Understand the importance of attending meetings and getting to know product and documentation development team members
   • Recognize the different types of documents that you might be required to test
   • Know how to choose the type of test most appropriate for the document being tested
   • Be able to develop a documentation testing plan
   • Know how to make any preparations necessary for testing product documentation

Attend meetings and get to know team members

The earliest preparation that can be made for documentation testing is that of joining the product and documentation development teams and becoming an active and useful member. A documentation tester has certain responsibilities to both the product and documentation teams. Some of these responsibilities are common to being a member of both teams including

- Attending meetings when scheduled
- Arriving on time to meetings
- Coming prepared to participate as specified in the meeting agenda

While these responsibilities are mostly common sense, they are also polite business practice. In addition, they are expected of not only documentation testers, but of all team members. Following common business practices of this nature is important to the alignment of the team. *Alignment* is defined as a group of people functioning as a whole. This alignment of team provides several benefits (Senge 1990) including

- Less wasted energy
- Commonality of direction
- Synergy
- Shared vision
- Understanding of how team member's efforts complement one another

When several individuals share responsibility for a single product, there is likely to be substantial duplication of effort. A team that emphasizes communication and work tracking can avoid much duplication of effort. Less duplication of effort also results in less wasted energy.

Few products are developed that could have been developed in only one way. There are many solutions to any problem. One of the benefits of being a team member is that you can contribute to the final choice of potential solutions. Once the team makes its choice, all members know and understand that decision, then work toward the same end. All team members have a commonality of direction. Attending team meetings helps to ensure that you have input into the final decision. It also helps to keep you working toward the same goal, following a common path.

Being a member of these teams is also important to the ultimate success of these teams, as well as to the teams' synergy—the ability of two or more (people) to accomplish more than what could be accomplished separately by each. A team that functions synergistically is far more productive than one that does not.

Synergistic teams also tend to share the same vision. This too tends to help the team be synergistic. It also helps to ensure that each team member follows the same path toward the same goal. Attending team meetings helps you to work toward accomplishing the same goals as the team. Attending team meetings lets you see how each of the chosen team members contributes to the success of the team, and to the success of the product being designed and developed. In addition, it also lets you to see how you fit into the team and the development cycle.

There are several benefits associated with team participation. There are also responsibilities associated with being a team member. Because the documentation tester is a member of two teams, the documentation tester has two sets of primary responsibilities, in addition to those already discussed and which are common to membership in both teams. The primary responsibilities of the documentation tester to the product development team include

- Gathering information and knowledge
- Providing input as appropriate
- Educating product development team members

Documentation testers gather information about various aspects of the product, including information about the products being developed and sold as the product's competitors. The information that the documentation tester gathers is useful to the product designers, developers, testers, and documentation writers, in addition to being useful to the documentation tester, and therefore, should be shared with the team.

Product development team members use the information the documentation tester and other team members provide to create an effective, high-quality product. Marketing documents such as the MRD, and practical, user-oriented input that the documentation tester and others provide, help the entire team design and create a better product.

The knowledge that a documentation tester gains of the product becomes very useful, particularly when it is time to conduct the actual documentation test. Already having a strong knowledge of the product makes it easier to determine which type of documentation test is most appropriate for the document. In addition, it makes it possible for the tester know where to be particularly diligent when looking for potential problem areas in the product as well as in the documentation.

There is also another benefit to the tester's increased product knowledge. Because the documentation tester is involved with the product from the start, the documentation tester's product knowledge is often more complete than any other single team member. The documentation tester has the opportunity to learn about all aspects of the product, good and bad. The result is that the tester often has more exposure to and knowledge of the product than any other individual team member.

Along with this knowledge comes another responsibility, that of providing relevant information to the product development team as needed. The documentation tester's overall view of the product and its competition puts the tester in the position of being able to offer suggestions and information that contribute to the product's development and increase its overall quality.

Not only must the documentation tester participate in and contribute to product development and product development team meetings, the tester must also work toward increasing team member knowledge. There are two areas in which the documentation tester is particularly qualified to help increase product development team member's knowledge: documentation development and product interface design.

The documentation tester is uniquely qualified to provide input and information related to the development of the product's accompanying documentation. The documentation tester fully understands the document development process, and can help answer related questions.

For example, a product development team trying to determine whether on-line or paper documentation would be more suitable for the particular software application program being designed could utilize related information from the documentation tester. The tester could answer questions as to which would be quicker to develop, or how much each will cost.

The documentation tester is also often uniquely qualified to provide the development team with information related to the product interface. Of all development team members, the documentation tester is generally the one individual who can see the product from the potential user's aspect, as well as from the developer's aspect. When these two aspects are in opposition, the documentation tester's knowledge and input can help the development team make the best choices.

The documentation tester has responsibilities to the documentation development team as well. The documentation tester has the same responsibilities to the documentation development team as he does to the product development team:

- Gathering information and knowledge
- Providing input as appropriate
- Educating documentation development team members

In addition, the documentation tester has other responsibilities to the document development team:

- Providing document reviews
- Participating in the assessment of related documentation

Documentation development teams often use some of their team meetings to begin developing or to review the contents of documents. The documentation writers often meet early in the project's development to determine what should go into the product documentation. The writers and other members of the documentation team discuss the audience, purpose, and basic content of each document. The documentation tester can participate in these early meetings, providing some of this needed information. This is particularly true in environments where other technical documentation, such as an MRD, is not available.

As the project moves forward, the documentation team meets and reviews such documentation-related items as a proposed table of contents or list of index words. While the documentation tester might not participate in the earliest of the planning meetings, the tester does begin to participate in documentation meetings once document outline review begins.

While important to the development of the documentation, the documentation tester does not usually review a document's table of contents or index during the normal course of a documentation test. Indexes and sometimes the final table of contents are generally prepared too close to the end of the documentation development cycle to be included in the actual test. Therefore, reviewing these and similar documents early in the development cycle contributes to an improved document in two ways.

First, reviewing these documents, or portions of documents, early in the document development cycle helps to ensure the documents are being designed to be as suitable and as useful as possible. Second, reviews of some document portions, particularly the list of index words, helps to ensure all portions of the document are reviewed, at least to some extent. This approach includes the portions of the documents that cannot be reviewed during the testing cycle because they are not complete.

A documentation tester might also participate in the assessment of related documentation. When more than one piece of documentation is prepared for a product, more than one documentation tester might be responsible for testing the product's documents. During team meetings, each documentation tester should have the opportunity to participate in the development or review of documentation plans for each of the product's related pieces of documentation. These team meetings might be the only time in which a documentation tester gets to review each of the pieces of documentation to be included in the set.

As part of participating in the assessment of related documentation, the documentation tester can also contribute to the development of documentation plans by participating in the early review of those plans. The documentation tester can help to pinpoint areas where the plan needs improvement or additions and make suggestions to provide these improvements or additions.

Documentation testing plans also provide much of the information a tester needs to know in order to prepare for a documentation test. Therefore, reading, reviewing, and participating in the development of documentation testing plans is an important step in planning and preparing for a documentation test. As such, it is also an important reason for being an active member of the documentation team.

Documentation plans

The documentation plan is the most important piece of written reference material the tester has from which to prepare for a documentation test. Other documents such as the MRD and PDR provide information about the user's needs and wants and the eventual design of the product, so they too are very useful. The documentation plan, however, shows the tester what types of documents will be created, as well as what information those documents will contain.

Of the information contained in the documentation plan, information about the types of documents the writer is planning to create lets the documentation tester know which types of tests will be most suitable for each document. It can be very difficult, if not impossible, to plan and prepare for a documentation test when the type of document to be tested is not known.

Just reading a documentation plan, however, does not necessarily provide the tester with all of the information that is needed. Understanding what the documentation plan should provide helps the tester know when the documentation plan is not providing that information. Therefore, the documentation tester needs to understand how a documentation plan is prepared, as well as what types of information it contains.

Documentation plan components

The documentation plan contains several important components. While the documentation plan might differ from one company to another, many components are standard or at least similar from company to company. The most common components of a documentation plan are

- Document description
- Target audience
- Document overview
- Document correlation
- Proposed table of contents
- Document synopsis
- Production dependencies
- Production information
- Documentation schedule

Document description

In most cases, some preliminary information is basic and necessary to understanding the information to follow. In a documentation plan, that preliminary information includes a proposed document title, a unique company identifier, and revision history.

The proposed document title should be as short and as descriptive as possible. "Concepts Manual," "Playset Installation Instructions," and "Users Guide to Networking" are all short and descriptive document titles.

The unique company identifier should identify the document as unique from any other document in the company. The Library of Congress, for example, uses ISBN (internal standard book number) numbers to identify all books it catalogs. No two books carry the same ISBN number. ISBN number prefixes are assigned to publishers around the world. Unless your business is a publishing house, however, you are not likely to use ISBN numbers on your publications (Huenefeld 1990). That does not, however, preclude your company from establishing a cataloging or numbering system for your internal publications. If such a system exists in your company, you need to determine the appropriate identifier to be used for the book or books included in the documentation plan.

This section of the documentation plan might also include a list of the members of the documentation team and others responsible for the document's production.

Target audience

The target audience section does not need to be very detailed because it is included in documents such as the MRD. Instead, this section can be a shortened version of an audience definition that is contained in other company documents, or which you or another documentation team member have created.

If no target audience description exists, or the audience has changed substantially from that described in the original research documentation, then you can write your own target audience description. At this point, a good understanding of your target audience is important to ensure the success of your document.

If details about your audience are not readily available, you need to seek out the information. One way of successfully determining and defining your audience is by asking questions about your audience. Figure 3-1 shows an audience worksheet (Woolever and Loeb 1994). This worksheet was developed specifically to determine the audience for computer manuals. This figure should, however, help the writer determine the types of questions that need to be asked. The results, as de-

---

AUDIENCE WORKSHEET
Primary Audience

1. What is the job function of the primary audience?
2. How will this audience use the document?
3. What is the educational level of this audience?
4. How experienced are the members of this audience in their jobs?
5. How experienced are they with computers? With your product?
6. What is their work environment like?
7. What is their interest level?
8. What biases, preferences, or expectations might they have?
9. With what other computer documentation are they familiar?
10. How much theory or "nice-to-know" information do they want?

3-1 *Audience worksheet*

scribed in this section of the documentation plan, help the documentation tester understand the audience the writer believes he is addressing.

Document overview

This section contains a brief description of the document being created. It explains the document's purpose and describes the needs or wants of the target audience. This section helps the documentation tester better understand the document's user as viewed by the writer. It gives the documentation tester the writer's point of view, from which the document will be written.

Document correlation

This section of the documentation plan explains how the document fits into the company's overall documentation set. If the document is not part of a set of company documents, its relationship to the competitor's documentation can be described instead. This information shows the entire documentation team, including the documentation tester, how this piece of documentation helps to fulfill the company's documentation requirements.

There is one other benefit of knowing the information contained in this section. In instances where a document is part of a set, this information helps the writer to ensure that all needed information is covered in the documentation set. By reviewing all documentation plans and comparing these sections, the documentation tester will have a general idea of what should be included in the document(s) for which he is responsible. This portion of each documentation plan helps the tester determine whether information that might be missing from the document being tested should be included in this or one of the other manuals in the set.

Proposed table of contents

The table of contents should list each chapter to be included in the document, as well as any appendixes and glossaries. If possible, the table of contents should be two or more levels deep to give the tester and other interested individuals as much detail about each chapter's content as possible. The tester can then use this information to determine what equipment or tools are needed for testing, as well as determining the most appropriate approach to testing.

Document synopsis

This section should include a brief description of each chapter. It should summarize in one or two paragraphs the type of information each chapter will include, as well as provide an idea of how the information in each chapter contributes to the overall success of the document.

Production dependencies

To modify an old saying—no man(ual) is an island. Neither is a writer, a tester, or any of the many individuals who contribute to the creation of a product and its documentation. All aspects of the creation and development of a product are intertwined. A problem in one aspect of production contributes to problems in other areas of production. These potential problem areas dictate the dependencies on which successful completion of the manual relies. This section of the documentation plan should spell out all known and potential dependencies.

For example, in software development, the creation of a user manual depends heavily on the completion of the design of interface screens. A writer cannot explain what choices are available to the user if the list of choices and their placement on the screen has not been determined by the software engineer. Information about this and other types of related dependencies should be included in this section of the documentation plan. Knowing the dependencies the writer has also helps the tester understand and plan for dependencies of his own. In addition, it helps the tester know what might interfere with the successful completion of the document test within the specified time limit.

Production information

The information contained in this section ranges from a simple description of the paper size, number of pages the document will contain, type of paper on which the document is to be printed, and so on, to a detailed description of every aspect of the production process. If the document is to be printed in-house (by the company's own printing department) more detailed information about the document is needed. Information such as the number of graphics, the type of graphics, how the document is to be bound, and other details should be included here.

In larger companies, document printing is often contracted to outside sources. In smaller companies, however, the actual production of the document might be done by the same people who write the document. In this case, you must supply more detailed information in the documentation plan, and you might need to work with your editor and with outside vendors to get the complete picture. The necessary information (Woolever and Loeb 1994) includes

- Approximate number of pages
- Type of production (that is, whether the document will be typeset, desktop-published, or produced by another method)

- Paper stock
- Typeface or font
- Use of color (one-, two-, three-, or four-color?)
- Type of binding (perfect-bound, spiral-bound or saddle-stitched?)
- Kinds of graphics (line drawings, schematics, screens, or charts?)
- Tabs
- Pullouts

For the most part, the documentation tester does not need to know this information. Its primary use to the tester is to give the tester an idea of the constraints under which the writer must develop the documentation. However, other members of the team do need this information for planning and production.

Documentation schedule

The last piece of information needed in the documentation testing plan is the schedule. This is particularly useful to the documentation tester. The schedule for testing the product documentation is significantly effected by the schedule for its creation.

Scheduling information can be presented in any of several formats. For example, it can be presented as a simple list with expected completion dates, as a form with task descriptions and due dates, or as a chart with time lines (Fig. 3-2).

All of the information contained in the documentation plan is useful to the documentation tester, to one degree or another. Some of the items are particularly useful in helping the documentation tester determine which type of test is the best choice for the document to be tested. Before you can make a choice of tests, however, it helps to know what test choices need to be made.

Choosing a documentation test

Once the documentation tester understands the documentation plan and its contents, the tester can then begin to determine which type of documentation test would be most suitable. Of course, the type of test chosen depends on several factors:

- The length of time available for the test
- The previous status of the document
- The type (or types) of documents being written

Length of time

The documentation schedule section of the documentation plan provides the information for determining the length of time available for

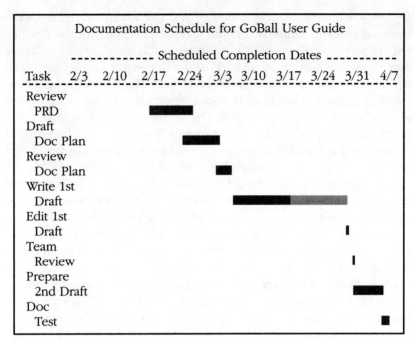

3-2 *Sample scheduling chart*

the test. It is only one item of input, however, and is not the sole determining factor. Just because the writer scheduled one week for a documentation test, for example, does not mean that the documentation tester can conduct the desired level of test in that time frame. It is the documentation tester's responsibility to negotiate the length of time available for the test. There might need to be some give and take for all parties concerned. The ultimate goal of improving the quality of the documentation should be a primary concern.

Whenever possible, the negotiation for the type and length of documentation test should be conducted during documentation team meetings. If accomplished as intended, the length of time available for the test to be conducted will be included in the documentation plan. The time shown on the plan will have been agreed on by the writer and the tester, and should not come as a surprise to either.

Previous document status

Whether the document is new or a revision of an existing document will affect the type of test conducted on the document. If the document is new, the test should be as extensive as possible, and should include all of the document. If the document is a revision of an earlier version, the type of test and what is tested might be less extensive. For example, if only some of the document's chapters were changed from the original version, it is not necessary for the remaining chapters to be tested. It is common in this situation to conduct a more complete test on those parts of the document that are new, and just do a simple read-through test on the remainder of the document.

This particular situation is also an example of one in which the tester might be especially useful. The writer can avoid reviewing the unchanged chapters, but the document's quality can still be ensured by having the documentation tester read the document to verify that nothing else should be changed. It is relatively easy to overlook a minor change or two that should be made when the writer is concentrating on the overall document. This type of error or omission can be found by the documentation tester in time to correct it.

The type of test conducted on the document originally, if the document is a revision, will also affect the type of test conducted this time around. If the document received only a minimal test such as a read-through when it was first created, a more exacting test might be called for on the revision. In this situation, even the unchanged chapters should be given a thorough test as appropriate for the type of document being tested.

Type of document

The type of document also affects the type of test to be conducted. For example, it simply is not possible to conduct a task-oriented test on a conceptual document. A read-through level test is more appropriate, and more effective, for a conceptual document than is any attempt at performing tasks or following a procedure, neither of which are included in a conceptual document.

The documentation plan explains the type of document being created. This in turn is used by the tester to determine the most suitable type of test to conduct. Of course, this means that the documentation tester needs to understand what types of documents can be created, what each type of document is intended to do for its audience, and other related information.

Types of documents

Although there are almost as many document variations as there are companies, there are three common categories or types of documents prepared for products:

- Conceptual
- Procedural
- Quick reference

Conceptual

Conceptual documents provide the reader with information about various aspects of the product that might not be included in other types of documents. A concepts document consists primarily of explanations of any or all aspects of the product. These explanations usually consist of two parts.

The first part of a conceptual explanation is generally provided as a brief definition. This is particularly true if the concept being discussed is normally referred to by an *acronym* (a word created using the first letters of a several terms, such as *WAC* for Women's Army Corp). The second part of a conceptual explanation provides detailed information about the term. Sometimes these detailed explanations are brief, but just as often, they can be very long, running into several pages.

In its most generic form, a concept document is an extended glossary of terms related to the product. For example, Novell, Inc., produces concepts manuals for many of their networking software products. Their concepts manual for their NetWare 4.0 network operating system product is called *NetWare 4.0 Concepts*. This manual presents its concepts in alphabetical order, ranging from *Abend*, "(Abnormal end) A message issued by the operating system when it detects a serious problem, such as a hardware or software failure. The abend stops the NetWare server. Abend messages are documented in *System Messages*.", to *Zones*, "Arbitrary groups of nodes on an AppleTalk internetwork. . . . See Also 'AppleTalk protocols.'"(Novell 1993)

Not all conceptual documents are separate documents. In many instances, conceptual information is contained within other types of documents. In *DOS Workstations End User Guide for NetWare 3*, conceptual information is presented at the beginning of the book under a section titled, "Words You Should Know." This section provides an alphabetized list of terms referred to in the rest of the book.

When provided as part of another book, conceptual information is generally reported in greater brevity than when a full book has been created to provide conceptual information. For example, in *NetWare 4.0 Concepts*, the term *rights* is described in detail, taking five and one-half pages to do so. On the other hand, in *DOS Workstations End User Guide for NetWare 3*, which provides network users with basic information they need to know in order to be successful at using the network and its resources, only one paragraph is devoted to describing the term *rights*.

When a documentation tester is preparing to test a manual, the tester must take into consideration the type of document to be tested.

When the document is of one type, planning and preparing for the documentation test is easier than when it consists of several document types. This book contains not only a section that is primarily conceptual, but also contains several sections that are procedural in nature.

Procedural

A document that is procedural in nature concentrates on providing instructions for the completion of a particular task or set of tasks. To again use *DOS Workstations End User Guide for NetWare 3* as an example, the remainder of the book concentrates on providing procedural information. It contains three sections, of which even the titles are procedural in nature:

- Common NetWare Tasks
- Printing Tasks
- Drive Mapping Tasks

These three sections are then further divided into procedures that describe how to perform these various tasks. For example, Fig. 3-3 shows page 13 from *DOS Workstations End User Guide for NetWare 3*. It contains the procedure to be followed if a network user wants to control whether or not messages are received at their workstation.

You should also note in that another form of providing conceptual information is used in this book as well (see Fig. 3-3). At the beginning of each procedure, a heading titled "Concept" is presented, followed by a brief description of the concept or purpose of the procedure that follows.

Quick reference

Brevity is the key to a quick reference document. They are usually written and designed for individuals who are very familiar with the product, but who might need occasional help with one aspect of it or another. This type of document might consist of a single page, several pages, or a small index card.

Quick reference documents are procedural in nature. They never include any kind of conceptual information. The procedural information they provide is generally sketchy compared to an actual procedure manual. A quick reference document often covers only the major or most important steps required to perform a given task.

Like conceptual and procedural documents, quick reference documents must also be tested to ensure their quality. One thing to keep in mind when testing this type of a document is whether or not its content is sufficient for its intended user. This is particularly important

Turning messages on and off	
Concept	You can control whether or not you receive messages.
Steps	To avoid receiving messages from other users. 1 At the DOS prompt, type castoff, then press Enter.
Helpful hint	Users who send messages to you while you have *castoff* activated will receive a message telling them you did not receive their message.
Steps	To start receiving messages again after blocking them with the *castoff* command. 1 At the DOS prompt type caston, then press Enter.
Other helpful topics	● Sending a message, page 12

3-3 *Sample page from* DOS Workstations End User Guide for NetWare 3

for a quick reference document because so much information is deliberately left out.

There are many variations and titles associated with these three types of documents. Regardless of the titles used, these documents can be broadly defined using these three categories. The category into which each document falls helps to determine the type of test that can and should be conducted on it.

One other factor that strongly influences the type of test to be conducted on a document is whether the document itself is intended to be delivered as printed or electronic documentation. The main difference is that of testing for functionality. If the document is to be delivered in electronic format, not only must the tester check the accuracy of the content, but the tester must also check the functionality of the electronics related to the document.

For example, most electronic documents contain *hyperlinks*—key words or icons that move you to another location in the electronic document. If a hyperlink is indicated in a document, the tester must ensure that the hyperlink functions, and that it moves the reader to the correct location in the document.

In addition, hyperlinks are generally provided that return the document to the original location from which the user executed the hyperlink. The tester must also ensure that the return works for each hyperlink in the document.

There is one exception to the tester being required to test hyperlinks. If the hyperlink testing within a document is part of the complete product testing, the documentation tester does not need to test the hyperlinks. More often than not, however, the product tester checks only the linking mechanism, not the individual links within a document. Even if the product tester does check each individual link, the check is to determine whether or not the link functions, not whether the link is made to the correct item of information.

Information about electronic document preparation is also provided in the documentation plan, and is one more reason why reviewing the documentation plan is important to the documentation tester. This and other related information should be considered when planning for and choosing a documentation test.

Choose the type of documentation test to perform

As previously explained, one of the major factors to be considered when choosing the type of documentation test to perform is what type of documentation is being prepared. Some types of documentation tests are more appropriate for different types of documents than are other types of tests.

For example, an engineering review or a read-through test is more appropriate for a conceptual document than is a usability test. Conceptual documents usually contain no procedures or tasks to perform. In this case, it would be almost impossible to perform a usability test. You might be able to conduct a usability test to determine whether the chosen layout for a conceptual document is simple and easy to use, but for the most part, this would be an insufficient test for this type of document. A more practical and useful approach would be to perform a read-through test to determine whether or not the quality and quantity of information is appropriate.

Some documents are a combination of types, containing both conceptual and procedural information. In this case, a combination of testing approaches might be needed. Read-through testing will probably be the best approach to take for the conceptual information contained throughout the book, while a basic functionality test might need to be done on the procedural portions of the book. In this case, the test to be planned and prepared for is generally the more detailed of the two—the basic functionality test—while the conceptual pieces will simply be given a read-through during the basic test.

Of course, choosing a test depends on other factors in addition to the type of document that will be tested. Being able to successfully choose the most appropriate documentation test depends on understanding the types of documentation tests that can be performed. The most common types of documentation tests are

- Read-through
- Engineering review
- Usability
- Basic functionality
- Integration

Read-through

The read-through documentation test is just what its name implies. It includes a reading of the document being tested. The primary purpose of the read-through test is to ensure that the information contained in the document is technically correct and complete.

There are several different types of read-through tests. Some companies already commonly conduct read-through tests, although they might not recognize or identify the process as being a read-through test. In addition, in companies without an individual filling the title of documentation tester, the type of read-through test performed is usually not quite the same as that performed by a documentation tester.

The most common type of read-through test is often called a peer review. A peer review consists of having one qualified individual read what has been written by another individual. They can be conducted on documents written by engineers, marketers, or other company personnel.

When a writer performs a peer review of another writer's document, he is looking for several items. Some of the items are the same as those the documentation tester would be looking for, and some are different. The most likely example of a difference is that of grammar and spelling. While a writer performing a peer review on a document would look heavily for proper grammar and spelling, a documentation tester would only note glaring errors. More often than not, a documentation tester will not find grammar and spelling errors, as the tester's concentration is on the technical accuracy of the document, not on whether everything has been properly punctuated.

Writers also look for consistency, document flow, proper formatting, and style, in addition to grammar and punctuation. A documentation tester is concerned with many of these issues as well. Certainly consistency and document flow are important to the quality of the document. In fact, consistency and document flow might also be important to the technical accuracy. Therefore, a documentation tester would place an emphasis on these two particular aspects of the document when reviewing it.

Even so, the primary purpose of conducting a read-through documentation test is to ensure that the document states all facts correctly, and that it includes all of the relevant information. For a documentation tester to determine whether or not the document is complete and correct, the documentation tester must have a great deal of knowledge about the product, because the tester must be able to detect errors and omissions without additional research.

This is not to insinuate that a documentation tester performing a read-through test does not do any research. That is far from true. Most of the time, however, the documentation tester relies on his knowledge when conducting a read-through test, and researches the answer to questions or the accuracy of a given statement only when he lacks sufficient knowledge in that area.

By its very nature, the read-through test is commonly applied to documents that are conceptual in design rather than procedural. In order to verify the accuracy of a procedure, the tester generally must conduct a different type of test, such as a basic functionality test.

Documents that have had limited changes made to them since their last release, and which were tested thoroughly at that time, can also be tested using a read-through test. Of course, this assumes that the changes are relatively minor in nature. Major changes, or changes to large sections of the document, also require a different type of test. For documents with extensive changes, a read-through level test is not generally sufficient, with one exception. If the document consists solely of conceptual information, a read-through test is then generally sufficient.

Engineering review

An engineering review test is similar to a read-through test. Most often this type of test is conducted by a product engineer, rather than by a documentation tester. The main exception is when the documentation tester is also a product engineer. This type of test can only be successfully conducted by an individual both the engineer and tester in reviewing a document written by someone other than himself. An engineer who is a documentation tester and the document's author cannot objectively and successfully conduct a satisfactory engineering review test on a document.

While very similar to the read-through test, the engineering review test can differ significantly in that the engineer has access to two other information verification items. Access to these additional sources of information often prove to be more limited for the documentation tester.

The first source of information is that of other engineers. The engineer/tester often has a closer relationship to other product engineers than the tester. In fact, engineers commonly sit in the same area, and can talk "over cubicle walls" without a great deal of difficulty. This is not generally a privilege afforded to the documentation tester. Quick answers to questions given to the engineer/tester by other engineers can speed the process of engineering review.

Second, while the documentation tester might have basic equipment set up with which to perform the specified level of test, the product engineer often has access to a greater depth and breadth of test and laboratory equipment. Whereas the documentation tester performing a test on a software manual might have the most common type of computer setup on which to test the document, the product engineer might have the top five most commonly used computer setups on which to test the document.

While there are advantages such as greater access to equipment and product engineers for the engineer/tester, there is one main disadvantage. That disadvantage is, while the engineer/tester might have access to other engineers and additional equipment, it is not likely that the engineer/tester has enough time to take advantage of this advantage. In this case, the engineering review test ends up being no more effective than a read-through test. And, in some cases, it is even less effective, particularly if the engineer performing the test is one who initially had limited participation in the product's design and development.

Usability

A usability test is primarily aimed at determining what aspects of the product's interface are particularly useful or particularly difficult for the intended user. A usability test is often conducted on the product, as well as on the product's documentation.

Many company's have separate teams that conduct usability tests. The documentation tester's responsibilities in this case generally fall into the area of helping to assess what aspects of the document should be tested. Often, usability tests for documentation are designed to find and correct those areas of a document that discuss aspects of the product with which the product's designers and developers are not totally comfortable.

Conducting a usability test can be expensive, time consuming, and very limited. Because it is also generally quite intensive, it is not usually possible to conduct a usability test on all portions of a document. Gathering individuals who are representative of the intended audience, conducting the test, and analyzing and reporting the results must usually be done on only a small portion of the document.

The result of this type of test is generally an indication of how effective the approach taken by the document as a whole is in meeting the user's needs and wants. To determine the success of a document's approach at meeting a user's needs and wants, the usability tester generally chooses a typical sample of the documentation, and after giving the user one or more tasks to perform, gathers information about how successful the user was in accomplishing those tasks.

The documentation tester might be the individual who performs the required tasks, or might simply be an observer, watching other user's perform the assigned tasks. The documentation tester might also be responsible for defining, setting up, and conducting the test when there is no other usability testing group within the company. If there is a formal usability testing group in the company, the documentation tester's responsibilities are likely to be more limited, but can range from helping to determine areas of the documentation to be tested to attending the tests as an observer, or anything in between.

Basic functionality

The basic functionality test assesses the accuracy of a procedural or task-oriented document. That does not mean, however, that a basic functionality test cannot be used on documents that also contain other types of information. The documentation tester frequently works with documents that contain several types of documentation. Some of these documents instruct users to perform specific tasks or steps, while at the same time providing them with the conceptual aspects of those tasks.

In the basic functionality type of test, the primary concern is the accuracy of the procedure being described, as well as its ease of use. Documents that concentrate on providing procedural information are the types of documents that should receive a basic functionality test.

The basic functionality test requires that the documentation tester walk through all procedures, performing each task in the order instructed. It also requires the documentation tester to perform each step exactly as documented. This sometimes proves difficult for a documentation tester. The main reason is that the tester is often so familiar with the product and the documentation that by the time the test is conducted, it is easy for the tester to fill in any missing or inaccurate information, instead of finding and noting it.

If the intended audience is as familiar with the product and its use as is the documentation tester, then this is not a problem. In fact, it can show that the document contains exactly what it should contain. Most often, however, a procedural document of this nature is written for less-experienced users. The documentation tester must pretend to know almost nothing about the product when conducting this type of test.

One other drawback associated with conducting this type of test is that it generally requires a great deal of planning and preparation, and the test itself is often very time consuming. In addition, it usually requires that the setup and preparation include obtaining the product and any associated equipment. In addition, the tester must make certain that when the testing environment is established, he sets it up so as to represent the most common or frequently encountered setup or environment.

If testing the documentation in only one environment is not sufficient, and it might not be when there are several environments in which the product is used, a more detailed test must be used. This more detailed documentation test is referred to as an integration test.

Integration

An integration test is one in which the documentation tester attempts to take into consideration every environment or situation in which the document and product could be used, and then tests the document accordingly. Often it is not possible to test a document within every possible environment. In that case, the integration test becomes more like a series of basic functionality tests. Only the most important environments are used for the documentation test, and cross-testing to ensure that one does not cause problems for the other might not be performed. Whenever possible, however, any areas in the documentation that, while perfectly acceptable for one environment cause problems in another, must be rewritten to accommodate all environments. Proving how that can be accomplished is one of the responsibilities of a documentation tester.

As noted, one major drawback of this type of test, besides its time-consuming nature, is that not only do you have to make sure that the documentation is correct for two or more separate environments, you must make certain that the instructions for one of those environments does not negatively impact the others. Therefore, the more environments you add, the more complex the test becomes, and the more knowledgeable and thorough the documentation tester must be.

Part of being thorough and ensuring that all requirements are met before, during, and after conducting a documentation test involves proper preparation of a documentation testing plan. That plan, once created, should be reviewed and approved by other involved individuals, particularly the document's author. Therefore, it is important to understand what should go into a documentation testing plan in order to ensure that it is complete, accurate, and effective.

Develop a documentation testing plan

Developing a documentation testing plan requires attention to several key factors:

- Writer input
- Level of test
- Time required to test the documents

As with the documentation plan, the documentation testing plan is prepared with input from many sources, the most important of which is the person who is writing the document you will be testing. Therefore, just as the documentation tester provides input to the document's writer when the writer is preparing the documentation plan, the writer should be given the opportunity to provide input to the documentation tester's plan.

Writer input

Writer input can come in many forms. The two most common forms are the documentation plan and a formal or informal review of the documentation testing plan by the writer.

The documentation plan is a primary piece of information from which the documentation tester develops the testing plan. As noted earlier in this chapter, the documentation plan contains several components important to the tester.

The document description tells the tester the type of document that is to be tested. In particular, this section tells the tester whether the document will be a conceptual, procedural, or quick reference.

This information limits the type of test the documentation tester can consider.

The target audience information provided in the documentation plan tells the tester who the writer believes the document is being written for. It also tells the documentation tester who he should keep in mind when conducting the actual test.

The document overview also provides the tester with a basic idea of the kinds of information to be contained in the document. This information can help the tester determine whether or not all relevant topics will be covered. It also helps the tester begin to understand how he has to set up the testing environment in order to conduct a successful test.

How this document relates to other documents in the set is particularly important to the tester. If the tester believes that this manual is not covering all of the needed information, he can verify that the information is covered elsewhere by knowing what other manuals will be provided.

The proposed table of contents gives the tester an idea of how detailed the document will be. It also provides the first clue as to how many pages the document will contain. While an initial table of contents does not include page numbers, experience will show the tester approximately how many pages are needed in order to successfully cover the material listed.

A document synopsis provides the tester with a testing focus. For example, if the synopsis indicates that the purpose of the book is to help the user get started using a particular product, the tester can focus on whether or not the document does in fact help the user get started with the product. In this case, the tester's testing environment can also be set up to ensure that the user can easily get started with the product.

Any production dependencies that effect the writer will also effect the tester. Therefore, many testing plans include the same, or almost the same, list of dependencies as does the documentation plan. The testing plan might include additional dependencies, however, as there might be some aspects of the test that are independent of the writer. For example, testers generally must set up a testing environment, while writers often use an environment developed for them by engineers.

Production information is important to the documentation tester only from the standpoint of how the final document is to be produced. The most important item of information is generally whether the document will be printed on paper (hard copy format) or will be produced for electronic retrieval (soft copy format). If the document is to be electronic, the test should be conducted on the electronic form. However, this is not always possible without interfering with production. If that is the case, the test might have to be conducted on a printed copy of the document, instead of on an electronic copy.

Of course, the documentation schedule effects the time frame for conducting the documentation test. The test time frame might already be spelled out in the documentation schedule. If this is the case, the tester needs to be certain that he agrees with the amount of time set aside for the documentation test, and that the type of test to be conducted can be conducted within the given time frame.

Input to the documentation testing plan might come from sources other than the writer including

- Management personnel
- Other documentation testers
- Previous documentation testing plans

In some companies, management personnel define the majority of the input for a documentation testing plan. The definition can be direct or indirect. Direct definition would include such things as the type of test that is to be conducted and when it is to be complete. In an ideal world, the documentation tester makes these decisions based on the information he receives. In the real world, however, these types of documentation testing plan inputs might be dictated by management, and internal or external forces that effect the product.

Input from other documentation testers can prove particularly useful to new testers. When you first begin testing documentation, it can be very difficult to judge how long it takes to conduct each of the various types of tests. After some experience, your time estimates will become more accurate. Until then, other experienced testers can provide you with some guidelines.

Level of test

The level of test to be conducted is a key factor in a documentation testing plan. The levels of testing available to a documentation tester are

- Read-through
- Engineering review
- Usability
- Basic functionality
- Integration

As discussed previously, the level of test to be conducted must be indicated in the documentation testing plan. When you include the level of test to be conducted, also include a one- or two-sentence description of the test. That way, everyone who reads the documentation testing plan will understand what type of test is to be conducted. This also gives people an opportunity to disagree and suggest alternatives.

Time required to test documents

Other testers, as well as your own personal experience, will help you calculate the amount of time required to test documents. In case you do not have other testers from which to obtain this information, you can use the following table to guesstimate reasonable testing time lengths.

Table 3-1. Guidelines for estimating testing time requirements

Document type	Test type	Pages per hour
Conceptual	Read-through	2–8
	Engineering review	2–8
	Usability	1–15
	Basic functionality	N/A
	Integration	N/A
Procedural	Read-through	2–8
	Engineering review	2–10
	Usability	1–15
	Basic functionality	1–6
	Integration	1–3
Quick reference	Read-through	1–10
	Engineering review	3–20
	Usability	1–15
	Basic functionality	1–15
	Integration	1–10

Note: The number of pages per hour is shown as a range. The smaller of the two numbers is the minimum amount that a moderately experienced tester should be able to accomplish, given an average complexity of the material. The larger of the two numbers reflects a reasonable maximum number of pages, given the same level of complexity of the material and experience of the tester. Individual results will vary, but these estimates can be used as a starting point from which to begin scheduling time for a documentation test.

N/A indicates that a test of this type is not performed on a document of this type.

As Table 3-1 indicates, the time frame for conducting any one type of test on any given type of document can be rather broad. For example, a read-through test of a quick reference document can take anywhere from 6 to 60 minutes to complete one page. The length of time it takes is effected by several factors:

- The amount of knowledge the tester has of the product
- The amount of experience the tester has as a tester
- The complexity of the information contained on a single page of documentation

A documentation tester who has been a member of the product development and documentation teams from early in the product's design and development should have a higher level of product knowledge than does a tester who joined either team after the product design was well under way. In addition, a tester who is testing a document for an existing product that is being updated might have a great deal of product knowledge already, assuming that the tester is familiar with the earlier version of this product. Of course, this also assumes that the opposite is true. For a tester who is not familiar with the earlier version of the product, it might be necessary to become familiar with both versions before the current version's documentation can be tested.

A documentation tester who is experienced at testing various types of documents might be able to test quicker than an inexperienced tester. Self-confidence can have an impact on the time required to conduct documentation tests.

The document's complexity is also an important factor. For most testers, a product with which they are familiar because of exposure to it or similar products in their daily life is less complex, and thus easier to test. For products with which the tester has limited daily experience might be far more complex. For example, documentation for a space shuttle would seem less complex to an aerospace engineer than it would to an auto mechanic.

Write the testing plan

Give serious thought to the testing plan when you write it. You will be expected to follow the plan you have written, published, and received approval on. Mistakes in the testing plan can result in major setbacks and disruptions to the production schedule. You want to be noticed for your successes, not remembered for your failures. Make certain, therefore, that your testing plans are accurate, and that they allow you sufficient time to perform the test as you have outlined and defined it in the testing plan.

Perhaps the best way to show you how to write a testing plan is to provide you with an example. Figure 3-4 shows a testing plan used

---

WKGroup v2.1 On-line Help Documentation Testing Plan

Date Prepared: 3 September 1994
Prepared By: Lee Michaels

AUDIENCE DEFINITION
   The target customer is any company with small groups of people who need networking.
   Also, small businesses and home-based businesses are the secondary audience.

TESTING ENVIRONMENT
   Software Included: All help screens activated using the F1 key.
   How Product Distributed: Through retail computer software stores. (Off the shelf.)
   Operating Systems Included: Compatible with DOS-based Operating Systems.

COMPETITION
   FanTasticLan, WorkgroupsPlus, and NoviceLan are the leading competitor's products.
   Competitive Terminology: No special terms not also used in WKGroup v2.1.

TESTING PARAMETERS
   Test Level: This version will receive only a read-through level test. The 2.0 version of this
   product received a basic functionality test, and no new screens have been added.
   Purpose: In addition to the soft-copy of the help file, a printed copy of all help screens
   will be available for testing. Comments, concerns, et cetera are to be documented on the
   printed copy of the screens. The writer will input all changes to the soft-copy.
   Style Guide: The standard WorkingGroup, Inc., style guide will be followed for all documents

OBJECTIVE
   To ensure that all changes made to the screens are technically correct, and reflect the
   actual changes made to the product. Pay particular attention to help screens related to
   communication interface as substantial change was made in this area.

RESOURCES AND DEPENDENCIES
   Writers and developers must meet their scheduled deadlines for the test to go as planned.

3-4 *Sample testing plan*

to conduct a test on documentation that accompanied a software product. (Names have been changed to protect the innocent.)

Notice that the categories of information and their contents shown in Fig. 3-4 are basic to documentation testing plans in general. They reflect the information about documentation testing plans that has been discussed in this chapter.

Also note that your documentation testing plans do not need to be exactly like the one shown in Fig. 3-4. The most important aspect of a documentation testing plan is that it provide its readers with the information they need to know about the planned documentation test. If you have additional information you feel should be added to your documentation testing plans, add it. If some of the information shown in Fig. 3-4 seems inappropriate for your testing plan, leave it out. Most importantly, make an effort to make your documentation testing plan fill your company's need for documentation testing information.

Test plan acceptance

After you have written your documentation testing plan, have it reviewed and approved by members of the documentation and development teams. It might not be necessary for all team members to approve the plan. You might only need approval signatures from the team leaders and the document's author.

Regardless of whose signatures you get, be certain to get some type of written approval. You can use any form your company requests. If your company does not have an approval form, you should create one of your own. Figure 3-5 shows a sample form you can use as a guideline when developing a form more suitable to your company and your needs.

3-5 *Sample approval form*

Make the necessary testing preparations

Once you have created the testing plan and received the necessary approvals, you must prepare for the test. If the test is several weeks or months down the road, your preparations can wait until the testing time draws closer. In most cases, however, you are likely to find that you have barely enough time to get prepared for the test. But prepared you must be.

Define your needs

The first step in being prepared to conduct the test is that of defining your needs. Most of the time, your needs define themselves as you prepare the documentation testing plan. However, if the documentation testing plan does not spell out exactly what equipment or product you need in order to conduct your test, you should make a list of what you need. Include on this list a resource reference. The resource reference should explain where and how you intend to obtain the needed equipment.

Obtain necessary equipment and product

Once you have listed the equipment you need to obtain, you must then obtain it. If it means purchasing equipment, allow yourself sufficient lead time to have your purchase request processed. If it means you must borrow equipment, make certain you arrange ahead of time for the equipment you will be borrowing, who you will be borrowing it from, and when you will be returning it.

To help ensure that everything is ready when you need it, keep in touch with anyone who is helping you obtain the necessary equipment. Be certain to follow up on purchase orders, and verify that the equipment you will be borrowing will be available when you need it.

Set up the testing area

Once you know what you need to set up for and conduct the test, set up an area in which to conduct the test. In some cases, you might simply use your own desk or work area. When large sizes or quantities of equipment are involved, however, you might need to arrange for a larger area in which to set up everything you need.

Once you have a place to conduct the test and have gathered the equipment and related resources, set everything up so that accessing it is a simple matter when it comes time to test the documentation.

Since nothing ever goes exactly the way we plan, plan for and prepare to set up your equipment and related resources at least one week before they are needed. Nothing will throw your schedule off quicker than not being able to get the required equipment up and running in time to conduct your test.

Summary

Although this chapter might give you the impression that documentation testing is tedious, time consuming, and boring, only the first two are generally true. Documentation testing is an exciting and fun profession. It is also a rewarding one.

Even though a great deal of time, education, experience, and effort go into preparing for a documentation test, completing a test provides you with a sense of personal accomplishment. Successfully completing a documentation test helps to ensure that your company is producing the highest-quality product documentation that time and resources allowed.

With each documentation test that you conduct, you will find that your knowledge and skills grow. Even if your early experiences with documentation testing are painful (and most are not), documentation testing grows on you. The more you do it, the more fun it becomes, and the more satisfied you will be each time you complete a test.

I find it to be very rewarding to receive a printed copy of a document that I was responsible for testing. It is very satisfying to be able to hold that document up, look at it, and say, "I tested this document, and it's better because I did."

References

Grant, Todd. 1995. *DOS Workstations End User Guide for NetWare 3*. Spanish Fork, Utah: WriteTech, Inc.

Hetzel, Bill. 1988. *The Complete Guide to Software Testing*, 2nd ed. Wellesley, Massachusetts: QED Information Sciences, Inc.

Huenefeld, John. 1990. *The Huenefeld Guide to Book Publishing*. Bedford, Massachusetts: Mills & Sanderson.

Myers, G. J. 1979. *The Art of Software Testing*. New York: John Wiley & Sons.

Novell, Inc. 1993. *NetWare 4.0 Concepts*. Provo, Utah: Novell, Inc.

Senge, Peter M. 1990. *The Fifth Discipline: The Art and Practice of the Learning Organization*. New York: Doubleday Currency.

Woolever, Kristin R., and Helen M. Loeb. 1994. *Writing for the Computer Industry*. Englewood Cliffs, New Jersey: Prentice Hall.

4

Conducting documentation tests

As a documentation tester, all your preparation, all your studying, all your education is done for one purpose—to conduct documentation tests. Reading the introduction and the first three chapters of this book are done for the same purpose. This chapter explains the types of documentation tests you can conduct, the types of documents you can conduct them on, and how to conduct each type of test. This chapter might well be the most important chapter of this book, if your goal is to learn how, when, where, and why to conduct documentation tests.

After reading this chapter, you will understand
- Basics to keep in mind when conducting a documentation test
- Testing electronic versus paper documentation
- How to conduct each type of documentation test
- Tracking results and noting problems

Basics to keep in mind when conducting a test

Regardless of the type of documentation test you are conducting, or the type of document on which you are conducting the test, there are four important things to keep in mind when conducting the test:
- Why you are conducting the test (technical accuracy)
- Whether all necessary information is included in the document (completeness of information)

- How useful the document is to its audience (usability)
- How accurate the document is (document clarity)

Technical accuracy

You conduct tests on documents to ensure the highest possible quality for that document. One of the most important measures of document quality is technical accuracy.

It is important that your product documentation not contain errors. People who read your product's documentation usually do so either because they cannot use the product without some type of instruction, or because they are experiencing a problem with the associated product and need some help. In the latter instance, a user is generally already frustrated and possibly angry. If they find a minor mistake or two (such as a misspelled word or improperly punctuated sentence), they will probably forgive and forget such a simple mistake. However, they are not as likely to forgive technical inaccuracies that have a significant impact on them. Not only are users unwilling to forgive important mistakes, but technical inaccuracies often cause them to give up on the product altogether, demand a refund for the product, or request support from the company's technical support lines.

Ensuring the technical accuracy of the document you are testing is, therefore, your most important responsibility. It should also be your first consideration when conducting a test.

Completeness of the information

Closely associated with ensuring technical accuracy is that of ensuring the completeness of the information. A document that contains no technical inaccuracies might be considered a perfect document. But if that document leaves out instructions or information the reader needs, regardless of how technically accurate it is, the user will not have a very high opinion of the document.

Your second most important task when conducting a documentation test, therefore, is to ensure that all information the user needs, and which should be included in the document, is included in the document. That does not mean that one document should contain everything the user could possible need to know. What it does mean is that, if the document is intended to contain a certain type and quantity of information, then you need to ensure that it does.

For example, if the document you are testing is a concepts manual, then all of the concepts related to the product and which the user might need to know should be included in the manual. There might be concepts that will not be included in the manual, and rightfully so. If some of the concepts are too advanced for the intended audience, then it is best not to include them in the document. If some of the concepts are related to one aspect of the product that is not otherwise discussed in the manual, then the conceptual information related to that aspect of the product should not be included.

Ensuring the completeness of any document that is tested is the second most-important item with which you should be concerned when testing a document. If the document is technically correct, and it contains all of the relevant information the user needs, the document should be one of high quality. This might not be completely true, however.

Usability of the document

In addition to being technically correct and sufficiently complete, a document must also be usable. A usable document is one in which the reader can readily find what he is looking for and follow or use that information once it has been found.

Determining the usability of a document can be the most difficult of the four important basics to keep in mind when conducting a documentation test. How do you as a documentation tester determine the usability of a document you are testing, unless you are conducting an actual usability test?

There are several approaches you can take to determine the usability of a document when the documentation test you are conducting is something other than a usability test. The approach you choose depends to some extent on the type of document you are testing. All approaches do have one thing in common, however. They all require that you have a good understanding of the document's intended audience.

The easiest approach to take is that of putting yourself into the user's shoes, at least mentally. When you begin your documentation test, start by picturing yourself as the typical document user. As far as possible, mentally duplicate the environment and attitude the user might be functioning in at the time he is using the document.

For example, assume the product whose documentation you are testing is a forklift, and further assume the type of document being tested is a procedure manual. The procedure manual's primary purpose is to instruct the forklift operator on how to perform basic maneuvers using this brand and model of forklift. Because you know the manual's intended audience is forklift operators with experience in forklift operation, but not necessarily with experience operating this particular forklift, you mentally prepare yourself to test this document by assuming that you are an experienced forklift operator who has never used this particular model of forklift. You might also mentally prepare yourself by picturing a typical place in which this forklift would be used, such as a warehouse or warehouse-like retail store.

Of course, being able to mentally prepare yourself for such a task might mean that you have driven a forklift in the past. Often the best documentation testers are those who have been a user of a similar product at some time in the past. Just as readily, however, you might simply have gained a feel for what the job requires by conducting end-user research. However you accomplish it, the point is that by mentally placing yourself in the shoes of the intended audience, you are better prepared to assess whether the document you are testing can readily be used by the document's intended audience. By pretending to be a forklift operator, you can look at the document from a forklift operator's perspective.

Another way to ensure the usability of a document is to perform some usability sampling tests in addition to the actual documentation test. A usability sampling test is one in which you choose a likely topic or procedure, then attempt to find the related information in the manual in much the same way as you would expect the user to find it.

Sometimes usability sampling is conducted as a matter of course during a documentation test. You might not even be aware that you are conducting a usability sampling test. For example, assume you are reading a procedural manual that instructs you to "choose and apply the appropriate rights to the selected user." You know what a user is because you are working on a procedure related to a user. However, the term *rights* is new to you. This is the first time you have seen the word used in this manner.

Your first reaction as a documentation tester is likely to be the same reaction the document's reader might have. You ask yourself what rights are available to be applied to a user. To answer this question, you conduct a usability sampling test. You pause your documentation test at this point, and begin searching the document for an answer to the question of what rights are available. You check the index for the word *rights*. You check the table of contents for a heading that includes the word *rights*. If neither approach points you to a place in the manual where you can find out what rights are available, you might begin to question the usability of the manual.

The best usability test of any manual is the ability of a user to find the information he needs. If too many searches of this nature end up in frustration and unanswered questions, either the document is incomplete, or its current arrangement and contents are not sufficiently usable. If the document does in fact include the information for which you were searching, but finding it is not quick and simple, the problem is likely to be the design or layout of the manual. The problem is one of usability rather than of accuracy or completeness.

Another approach to verifying the usability of a document is that of conducting spot usability checks. Spot checks can be conducted by the assigned documentation tester or they can be conducted by other documentation testers and volunteers under the supervision of the documentation tester.

A spot usability test consists of listing five or six representative tasks, if it is a procedural document, or representative concepts, if it is a conceptual document, and conducting a brief usability test using these representative tasks or concepts. The documentation tester defines what is to be found, sets up a short usability test, then conducts the test and records the results. If most of the items being researched are found, and most of the testers participating in the spot usability test are successful, it is likely that the document is reasonably usable.

Document clarity

Once you have verified that the document is technically correct, it contains all of the relevant information the user needs, and it seems to be usable, the final item to keep in mind is whether the document is clear.

Assessing document clarity is not quite the same as ensuring its usability, although at first glance that might seem to be. A usable document is one whose contents and information are easy to find and easy to use once found. A document that is clear is one in which the style, format, and content of the document all work together to successfully relate the information it is designed to provide. In other words, the document is clear and easy to understand.

A document containing complete and accurate information about a product is usable only if that information can be readily accessed when needed. Even if the complete and accurate information can be readily accessed, it is only useful if it is also easy to understand—designed and written to provide the information clearly and easily.

Successful testing of any document, regardless of the type of document or the method of testing used, must include checking the document for these four general items—accuracy, completeness, usability, and clarity. These four items should be incorporated into each documentation test, although the tests for these four items can be incorporated to varying degrees.

Electronic versus paper documentation

It does not matter whether you are testing electronic documentation or printed documentation. The same testing procedures and approaches apply, although some variations must be adopted because there are differences between electronic and printed (paper) documentation.

Differences between electronic and paper documentation

The main difference between electronic and paper documentation is rather obvious. Paper documentation is documentation whose final production includes printing of the document on paper, also called hard-copy format. Electronic documentation is documentation whose final production includes delivery of the document in an electronic form, also called soft-copy format.

The most common use of electronic documentation is in the delivery system of products that are also primarily electronic, particularly computer software. Products whose main access is from an electronic device, particularly a computer, readily lend themselves to providing the associated documentation in the same format.

The main difference between electronic and paper documentation for a documentation tester involves the actual presentation of the information, and might involve the contents of the document as well.

Many professionals debate the merits of producing electronic documentation. There are two leading schools of thought related to electronic documentation. The first school believes that it is acceptable and suitable to place a copy of a manual originally designed for print into an electronic format, without altering the contents of that document. The second school believes that the presentation and value of the document is greatly reduced by not altering that document to take advantage of and be more suitable for electronic use.

Electronic documentation can take advantage of something commonly referred to as hyperlinks. A *hyperlink* is a key word or graphic within a document that, when activated, lets you jump from one location in the document to another location, either within the same document or within another electronic document. The paper equivalent of a hyperlink is the index. If you want to find more information about a topic within a manual you are referencing, the index is a logical place to look for other references related to that topic. There are several drawbacks associated with this method of information location in a hard-copy document not also associated with using hyperlinks in an electronic document.

The first problem is the physical one. Once you move from the section of the document you are currently reading to the index to find additional references, it is easy to lose your place in the document. You might find it more time consuming to attempt to return to your original location in the document than it is worth. In an electronic document, returning to your original location is often as simple as choosing "back" or a related menu item from an electronic menu. The electronic manual remembers your last location and returns or sends you "back" to that location when you request it to do so.

Another problem associated with using hard-copy documentation is that the references are more limited. In hard-copy documentation, the number of index entries that can be placed into a book are limited by the number of pages the publisher is willing to dedicate to an index. In addition, guidelines for producing index lists for written documentation often specify or suggest a limit to the number of index entries. In an electronic document, it is possible to create hyperlinks to every significant word in the manual, if the author chooses to do so.

Another drawback associated with an index in a hard-copy manual is that of referencing the topics covered in different manuals. In electronic documentation, any book in the electronic set can be referenced. Hyperlinks between electronic manuals make it possible to move from one subject or topic in one manual to the same or a completely different subject or topic within another electronic manual. To accomplish a similar task with hard-copy documentation generally requires a separate index, and access to the entire set of documentation, which is not always readily available.

Electronic documentation also provides what is known as context-sensitive help. When you are working with a software program and you find yourself unable to continue and in need of help, context-sensitive help lets you bring up information specific to the task or option you are working with at the time.

For example, if you are working with an electronic spreadsheet and you want to have the spreadsheet calculate a figure for you but you cannot remember how, you can request context-sensitive help. The related help screen will open and instructions for calculating the figure are provided. To accomplish the same thing with hard-copy documentation, you have to reference the table of contents or index, search for a topic heading or key word that represents the information you need, then thumb through the document until you locate the required information. With the electronic version of the manual, the entire process takes a few seconds or less to complete. With the hard-copy of the manual, the process could take several minutes. In addition, the result of the hard-copy search might prove unsuccessful, whereas the result of the soft-copy search will be related directly to the topic at hand if the electronic help is context sensitive.

There are other differences between electronic and paper documentation that affect how a documentation test is to be conducted on them, but these are the most important ones for the documentation tester. The documentation tester needs to be concerned not only with the content of electronic documentation, but also with its functionality.

Types of electronic documentation

The functionality of electronic documentation is affected by what type of electronic documentation it is. There are two main types of electronic documentation. Both have already been mentioned, although only one type—context-sensitive help—has been defined. The other common type of electronic documentation is known as the *electronic book*.

Electronic books

An electronic book is just what its name implies. It is a book put into electronic format for simplified and speedier recovery and access. Electronic books are sometimes designed to look like books. The idea is to make them easy to use by making them resemble and, to some extent, function like their hard-copy relative, the printed document.

Electronic books often contain a table of contents, an index, chapters, graphics (drawings or photographs), lists of figures, lists of tables, and other standard features common to printed books. Pages in the book are displayed on the screen, usually one page at a time. The page might scroll or the electronic page might be designed so that a single page of the electronic book is represented in one screen. Figure 4-1 is a sample of an electronic book page.

One difference between electronic books and printed paper books is that you can only access the information contained within the electronic book from a computer. The main difference is, however, the speed with which information can be found in the electronic book versus the paper book and the extensive linking of information within the electronic book. It is almost impossible to duplicate in a hard-copy book the depth and breadth of cross-referencing available in electronic documentation.

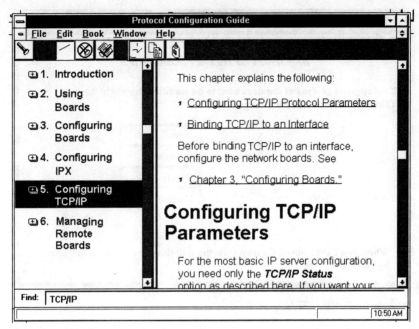

4-1 *Sample electronic book page*

Context-sensitive help

Context-sensitive help is accessed from inside a software program. You are typically already functioning from somewhere within the software program when help is required. The particular help made available to you is determined by the software, and is based on your location within the program or the function you are performing with the software. On-line help typically resembles the on-line help screen shown in Fig. 4-2.

Most commonly, you choose one of two options to start context-sensitive help. You either choose a predefined key on the keyboard, function key 1 (F1) and function key 3 (F3) are commonly defined for this purpose within various software programs, or you choose a button representing help. Most commonly the button says Help or uses a symbol such as a question mark (?) to represent help.

Buttons are often used when the software program is run using *buttons* (pictures resembling keys on a keyboard but with drawings or icons on them as representative symbols). Predefined keyboard keys, such as F1 or F3, are often used when the software program is run from the computer operating system prompt, most commonly from DOS.

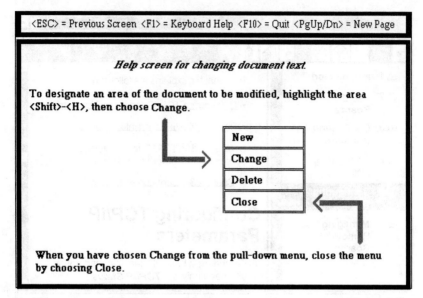

4-2 *Sample on-line help screen*

Context-sensitive help can be extensive, but is often more confined than the electronic book. Its very nature of providing help when and where you need it is also a constraint under which context-sensitive help must operate. So too is the platform on which it runs. For example, context-sensitive help activated by pressing the F1 or F3 key on a keyboard generally means that the program is a DOS (disk operating system) or other similarly based program. Because of the platform on which it was created, it is limited in its abilities. It can only provide functionality that the platform can support.

What to test for

The type of help provided, in addition to the information the document contains, determines what you as a documentation tester must test for in any given document. What you test for in an electronic document will be different than what you test for in a printed document. In addition, what you test for in context-sensitive help will differ from what you test for in electronic books.

Of course, there are some things you test for in all documentation whether it is presented in printed, context-sensitive, or electronic book format. These items—technical accuracy, completeness of information, usability, and document clarity—have already been discussed in detail.

In electronic text, as in printed documents, you check not only the content of the document, but also any references and cross-references. In printed documentation, checking the cross-references generally consists of verifying that the table of contents contains the correct page numbers, that index references are complete and point to the correct page numbers, and that references to other documents contained within the text of the document are correctly reflected. In electronic documents, you check cross-references in a slightly different manner.

In electronic books, cross-references are all linked. That is, you choose the cross-reference, and the computer jumps through the document to the information to which it is linked. Cross-references can include almost every single word in an electronic book, if the author chose to set it up that way. Usually only select words are chosen, but that can still account for hundreds or even thousands of cross-references to be checked.

In context-sensitive help, cross-references exist as separate chunks of documents. They are retrieved when requested. The information contained within them is separate from the software program, as well as from any related electronic documentation. Therefore, the contents of these help screens have to be read and tested just as printed documentation would be.

In this manner, electronic books and context-sensitive help differ. In electronic books, the link is made to other areas of the electronic book that contain related information. In context-sensitive help, the link is made to a separate document whose contents must be tested in addition to the other documentation.

In context-sensitive help, you are verifying that the right help screen opens, and that the information it contains is technically accurate. It is a two-step documentation testing process. On the other hand, in electronic books, you verify that the link opens, and that it opens to the correct area of the electronic book. The contents of the book that open when the link is activated do not have to be tested separately from the rest of the book. Because it is already part of the book, the information that is presented when the link is activated will be tested for technical accuracy, completeness, usability, and clarity when you test that part of the book.

How to conduct each type of documentation test

Technical accuracy, completeness, usability, and clarity are all key items to be tested for when you conduct a documentation test. The type of test you conduct determines how much time and effort goes into testing a document for these four items which contribute substantially to the quality of the document.

As discussed earlier, there are five basic types of documentation tests you can conduct: read-through, engineering review, usability, basic functionality, and integration. These five tests and how to conduct each one are discussed in detail in the following pages.

Read-through test

As the name implies, the read-through test consists of reading the document being tested. What the name does not imply, however, is the depth and effort associated with a read-through test. A read-through test is more than just sitting down and reading the document as if you were curling up with a good book by the fireplace. A better description of a read-through test might be to liken it to cramming for an examination when you have not yet read the book.

Even the analogy of cramming for a test as a parallel to conducting a read-through documentation test is not sufficient to describe what is actually required and what must be done to perform a successful read-through documentation test. When you are unfamiliar with potential test material, cramming requires you to absorb as much information as possible in as short a length of time as possible. This does require great effort, intensity of purpose, and concentration. These traits are also applied when conducting a read-through documentation test.

The biggest difference between cramming for a test and conducting a read-through test is that in cramming for a test, you are trying to absorb knowledge. In conducting a read-through documentation test, you must already have that knowledge and be capable of drawing on it to verify the technical accuracy of the information you are reading. The effort, intensity of purpose, and concentration required for cramming for a test and conducting a documentation test are relatively parallel, but they work to accomplish different ends.

As with all other types of documentation tests, you must prepare to conduct a read-through test. Some of the preparations you make for a read-through test will be different than for other types of tests, such as the basic functionality test. For example, in a read-through test you do not conduct any physical tests related to the document such as you would if the test were a basic functionality test that consisted of installing computerized software or setting up a children's play set. Even so, you still need to make certain preparations (Fig. 4-3), including establishing an area where you can comfortably conduct the test, if such an area does not already exist, and setting up the document for testing.

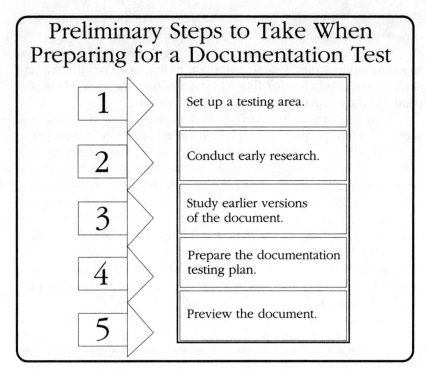

4-3 *Steps for preparing to conduct a documentation test*

Establishing an area where you can comfortably conduct the test is one of the first preparations you need to make. Most often, sitting at your desk is not the best place to conduct a documentation test, especially a read-through test. There are too many distractions—telephones, electronic mail, radios, and so on—that will interfere with your ability to concentrate. In addition, people who walk past your cubicle or come to your office and see you sitting and reading will assume you are reading because you have nothing better to do. They might see this as a sign that they should stop and chat. Or worse yet, your boss might get the impression that you do not have enough work to do and see to it that the situation is corrected. Therefore, even when you are conducting only a read-through test, you need to set up an area where the test can be effectively conducted.

You also need to do all of the same basic research that you would do for any other type of test. You need to attend documentation and development team meetings, and read and understand the MRD, PRD, documentation plan, and any other documents related to the document you will be testing. Whenever possible, you should get some practical, hands-on experience with the product. If the product is a play set, for example, you should try to assemble one yourself or observe others doing so.

If previous versions of the document exist, you should read them to get an understanding of the product. In this case, you are reading similar documentation for the purpose of learning more about the product whose documentation you will be testing.

Of course, as with all other types of documentation tests, you will write a documentation testing plan and get it approved. Just because you will only be reading through the document does not mean the test should not be conducted with the same professionalism as any other test. In addition, the test must be scheduled, all interested parties must be informed of what test will be conducted and how it will be conducted, and approval for the chosen type of test must also be given by the responsible parties. Preparing a documentation testing plan is the best way to accomplish all of these tasks. For these reasons, as well as to provide you with a guide and point of reference when you are conducting the test, you need to create a testing plan for the document.

You also need to preview as much of the document as you can. You will be trying to determine what areas of the document have changed, if the document is a revision to a previous version, and which areas of the document are completely new. In addition, if there are any references made within the document to other documents, you need to know that ahead of time. This allows you sufficient time to obtain a copy of any document referenced within the document you are testing, so you can verify those references as you conduct the test.

Previewing the document also provides you the opportunity to learn which aspects of the product are the least familiar to you. If you know the areas in your knowledge where holes or weaknesses exist, you have some lead time in which to fill in those holes, strengthen those weaknesses, or at the very least, find out who will be able to best answer any questions you might have when it comes time to conduct the test.

As appropriate, make any other preparations for conducting the test. You might find that making all of these preparations takes as long as, if not longer than, conducting the actual test.

Once the time to conduct the test arrives, you will be responsible for performing several tasks as part of conducting a read-through documentation test. The first task you perform is that of setting up the document so that it is ready to be tested. It is not uncommon to find that the writer has given you a document with no page numbers. In that case, the first thing you need to do is number the pages. Simply start with the number one (1) and write in the page number for each consecutive page. Where you put the page number and what it looks like does not really matter. What does matter is that when you start turning pages over, flipping back to previous pages, pulling out and setting aside pages with information you must check out later, jumping forward to check a reference, and so on, you can once again reassemble the document in its correct sequence.

Other preparations to be made on this document include obtaining sticky notes and colored pens with which to mark and flag changes and changed pages. You can use whatever works best for you when flagging changed pages. Post-It flag tapes have proven particularly useful to me because they come in several colors (good for flagging changes made by different documentation testers, or flagging changes made during different reviews of the same document), and are easy to apply and remove.

You might also find it easier to work with the document if you three-hole punch it and place it in a binder. The document pages that you receive for the test are generally printed on standard paper (8½ by 11), even though the final document will be printed in a different size. This is often true even when the final document will be distributed in electronic format.

Context-sensitive help screens are usually provided to the documentation tester printed on paper, one or two to a sheet. This saves paper (and trees), but makes it more difficult for the documentation tester to find the appropriate printed help screen to test it. It does, however, ensure that you see, read, and test all help screens. If you look at only the electronic version of the help screens, there might be some help screens that don't open. Having a printed copy of the help screens also provides a checklist for you to verify that you have in fact tested all of the related help screens.

Another setup technique that helps to speed the process when you begin the actual documentation test, particularly of context-sensitive help screens, is to prepare your own document table of contents, complete with the page numbers you wrote in earlier. For documents designed in book format, include the second- and third-level subheadings in the table of contents. For context-sensitive help, you need only include the title of each help screen. (Of course, this assumes that each help screen has a title, something which the writer should not have overlooked, and which you as the documentation tester should probably recommend if it was overlooked.)

Once you have completed all of this setup and preparation work, you will finally be ready to conduct the actual test. After training many different temporary employees as well as regular but inexperienced documentation testers to conduct this and other types of documentation tests, I have found that the best way to teach someone how to do a documentation test is to give them a portion of a document and walk them through the process. That is what the majority of the remainder of this chapter will do for you—show you samples of documentation and walk you through the process of conducting each type of documentation test.

Figure 4-4 shows a sample page from a user's guide for a coffeemaker. In conducting a read-through test on this page of documentation, you should look for and approach the test as described below and in the following pages.

Cleaning Your Coffeemaker   - 11 -

When to Clean Your Coffemaker

Clean your coffemaker whenever either of the following conditions occur:
- There is still water remaining in the reservoir after brewing a pot.
- Excessive steam escapes during the brew cycle.
- You have used the coffemaker daily for the past 30 days.

How to Clean Your Coffemaker

1. Wash the filter basket, water reservoir, and coffee pot in warm, sudsy water. Rinse thoroughly.
2. Place 1 cup white vinegar and 9 cups water into the water reservoir.
3. Add a paper filter to the filter basket.
4. Move the START switch to the ON position.
5. Stop the brewing cycle when 4 cups of vinegar water remain in the reservoir.
6. Wait 15 minutes then restart the brewing cycle.
7. When the brewing cycle is complete, dump the vinegar water.
8. Fill the reservoir with clear water, brew, then dump and rinse the pot.

4-4 *Sample page from a coffeemaker user's guide*

To conduct a read-through test on this type of a document, you must first become very familiar with the coffeemaker. Even if you are not a coffee drinker, you should still learn how to operate this particular coffeemaker. You need to understand how it operates, where the start button is located, what the filter basket is and how to remove it, how to fill the water reservoir, and so on.

Now, start at the top of the page. Notice the page title and page number. Check the page title against the table of contents (if provided) to verify that both headings are worded and spelled exactly the same way. Also, verify that the page number is correctly specified in the table of contents.

In addition, if the index is also supplied to you, note any words in the document's title or its contents that should be referenced in the index. Check the index to see if these are included. For this particular page (Fig. 4-4), words that probably should be included in the index are

- Cleaning
- Steam, excessive
- Water, remaining in reservoir

Continue the read-through by checking any subheadings that should be included in the table of contents. As with the section title, these subheadings should be properly spelled and written in the table of contents in the same way as they are written on the page. The page number references, if given, should also be correct.

If this page contained a graphic, now would be a good time to examine the graphic and, based on your knowledge, verify that it is accurate. You should also check the graphic's figure title for errors. The figure number should be checked and should follow the company's figure numbering style guidelines. In addition, the title should be correctly capitalized and all words correctly spelled.

Now that you have given the page a preliminary check, you should read it through. Start at the top of the page and read each sentence carefully to determine if it is technically correct, complete, usable, and clear.

For example, the first sentence reads, "Clean your coffeemaker whenever either of the following conditions occur:." This sentence appears to be correct, usable, and clear. It also seems to be complete once you finish it with the bulleted items. Therefore, you continue reading beginning with the first bulleted item which states, "There is still water remaining in the reservoir after brewing a pot." This bulleted item appears to successfully complete the first sentence, and continues to be correct, usable, and clear. After reading all three bullets you would probably believe that this particular section of the page is reasonably complete. It specifies when the user needs to clean the coffeemaker, and makes it easy for the user to find that information.

Did you notice the small problem with this section? The first sentence contains the word *either*. This word is associated with giving the reader a choice between two items. In this case, however, the bullet list contains three choices. The use of the word either in this particular instance is not technically accurate. As a documentation tester, it is your responsibility to note this discrepancy. The easiest way to do this is to circle the word *either*, and write a brief note near it. For example, you might simply say, "shouldn't this be 'any' rather than 'either'?"

As you have now noted something on this page for the writer's attention, this is also a good time to mark or flag the page to ensure that the writer finds and responds to your note. This might seem like a waste of time since this particular user's guide only contains about 16 pages. If this is the only page you note a correction on, however, the writer does not even have to look at the other 15 pages if you flag this page. This becomes particularly worthwhile when you work with larger documents, and is a courtesy the writer will appreciate.

Continuing with the read-through test, you read on to the next section and begin reading each numbered step. As you read, look for some of the types of mistakes that are easy to make when under a tight deadline including

- Missing a number or misnumbering one or more steps
- Not punctuating the steps with consistency
- Improperly spacing one or more of the steps

While it is true that an editor generally finds these types of mistakes during one of the edits conducted on the document, a good documentation tester points these types of problems out to the writer as well. A good documentation tester is also careful just to note the problem without making a big fuss over it, since it is not uncommon for the documentation tester to receive a document to be tested at the same time the editor is receiving a copy to be edited.

Read carefully through each listed step, and attempt to see yourself performing each step. Picture the coffeemaker in your mind. As you read step 1, see yourself washing the filter basket, the water reservoir, and the coffeepot in warm, sudsy water, then rinsing each of them. You will probably not have any problem seeing yourself completing each portion of this step.

But wait a minute. What happens if, when you picture yourself washing the filter basket, you also see yourself placing it in the dishwasher with tonight's dinner dishes? Did you just ruin the filter basket, or is it dishwasher safe? As a documentation tester, you probably know the answer to this question. Of course it is dishwasher safe. But how does the product's user know it is dishwasher safe? Perhaps it says it is on the outside of the coffeemaker's shipping box. Did the user read that information? Does the user remember that information? Should a note be added here, for the user's convenience, to remind the user that he can put some of the coffeemaker's parts into the dishwasher?

Well, you probably get the point. You should try to think like the end user. This is particularly important when you are conducting a read-through test, because you often cannot take the time to verify or obtain answers to every little question you might have. In this particular instance, I would probably write a note next to this sentence (most likely in the document's margin) asking the writer if he wants to make a note here that the filter basket, water reservoir, and coffeepot are dishwasher safe. I would then leave it at that. While I want to help the writer put in the correct information when what exists is not correct, I do not want to write the document for the writer. That is not a documentation tester's job, and will usually be resented by the writer, instead of accepted for what it is—an attempt to help the writer to create the best possible document.

With step 1 out of the way, continue reading through the other steps, noting any questions you have about the steps as you go along. Once you finish reading all of the steps, stop and ask yourself whether the steps were ordered properly, and if all relevant steps were included.

For this particular page of the document, I believe that all steps were included and that they were in the correct order. I had only one question when I finished reading the steps. I wondered if the reader should be told to discard the paper filter. After all, if he does not throw it away, he will be reusing it the next time he makes a pot of coffee. The vinegar in the filter paper might impart a strange taste to the coffee. As the documentation tester, I chose to add the following to the end of step 8, ", and discard the paper filter." The tested document page would now look much like the one shown in Fig. 4-5.

You conduct the remainder of the read-through documentation test in the same manner as just described. Continue to look for the types of problems and issues described here, as well as for any problems or issues that are unique to the type of document you are testing or to the product it was created for.

When you have completed your read-through documentation test, prepare the related and necessary reports. Reporting testing results is only minimally affected by the type of test you conduct. Therefore, report preparation is basically the same no matter which type of test you conduct. Reporting is discussed as a separate topic later in this chapter.

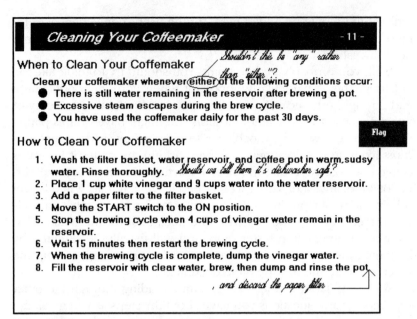

4-5 *Tested sample page from a coffeemaker user's guide*

Engineering review test

The engineering review test is very similar to the read-through test in that it most often involves reading the documentation for technical accuracy. The main difference is in the level of technical knowledge the documentation tester has about the product and its related documentation.

More often than not, an engineering review test is conducted on documentation that is very technical in nature. This type of documentation might never leave the company, because it is often intended for the use of those who are designing and developing the product and its related end-user documentation instead of for the actual end user. However, engineering review tests are sometimes conducted on documentation to be released to the product's intended audience.

Figure 4-6 shows a section of a page from a book that describes chelation therapy (Trowbridge and Walker 1991, 31). In conducting an engineering review test on this page of documentation, you should look for and approach the test as described below and in the following pages.

As with the read-through test, you should be looking for the same types of potential problems when conducting an engineering review

There are approximately 75 trillion cells in the human body, and each of these cells needs oxygen (among other elements) to function. When you receive intravenous chelation therapy--or even ingest oral chelating agents--you are helping all the cells utilize their oxygen. Nutritional scientists who administer the treatments do not know exactly how IV and oral chelators work. Only a few reactions are known relative to the vast numbers that exist, and the scientists are still studying them.

What is known is that since every body cell needs oxygen to function, by assisting in the oxygenation process, IV EDTA and oral chelating agents can, and do, significantly affect health and health-related problems, especially those involving the nerves, sex organs, and the cardiovascular system.

4-6 *Sample document page before engineering review*

test. For example, you must still verify that the information is technically accurate, complete, usable, and clear. You should also ensure that numbered items are correctly numbered, that section headings are listed in the table of contents in exactly the same way they are presented in the document, and so on.

In an engineering review, there are two very important aspects of the document on which you should concentrate. First is the technical accuracy of the document. An engineering review is often chosen as the type of test to conduct because it is the best-suited for very technical information. Because of the technical nature of the text, a simple read-through test is not sufficient. In addition, highly technical documents are likely to be more conceptual than procedural in nature. Therefore, a basic functionality or integration test is not appropriate.

Technical material requires that it be read for accuracy, above all else. However, the second important aspect of conducting an engineering review test is that of making certain the technical information is provided in such a way as to be clear and readily understood.

The two paragraphs shown in Fig. 4-6 contain technical references. The first reference comes in the first sentence. It states that, "There are approximately 75 trillion cells in the human body. . . ." This statement is made as a statement of fact. As such, its accuracy needs to be verified. If you are a documentation tester specializing in the health care or related field, you might already know this number to be technically accurate. If so, your knowledge is sufficient to verify this information. If, however, you are not sure that 75 trillion is the correct number, you must verify the number.

Verification of facts in technical documents can be done in one of two ways. First, you can find an authoritative source (or several if the number must be absolute) to verify the accuracy of the fact. If your source verifies the fact, you might be able to consider that good enough and go on. If your source disagrees with the stated fact, additional sources have to be found or an alternate method of verifying the information is needed.

The second way in which you can verify the accuracy of technical information is by conducting independent research. Most fields have published information that can be researched and used as a point of reference.

If you use either of these methods for verifying the accuracy of technical information, note the source you use. If you spoke with a specialist in the field, note who you spoke to, when you spoke to him, and what the results were of your conversation. Then make a note on the document for your own future reference, so that you can retrace your steps to report how you verified a fact or statement.

For example, in Fig. 4-6 you might begin your research into the approximate number of cells in the human body in an anatomy or related reference book such as *Gray's Anatomy* (Williams and Warwick 1980). If you find the information you are looking for in this reference book, then you might note it as "Grays" on the document being tested, placing additional details about where you found the information in a separate document.

One common problem that reoccurs in technical documents is the use of abbreviations, acronyms, and so on, without a prior introduction. When testing a document of this nature, be watchful for such occurrences. In Fig. 4-6, for example, the terms IV and EDTA are used. In this case, EDTA had been previously defined. Assuming that IV had not been previously defined, you as the documentation tester would insert the correct full name for this abbreviation. In this case, you would add "(intravenous)" to the document. Figure 4-7 shows the same document as was shown in Fig. 4-6, but with the documentation tester's two notes as previously discussed.

As with the read-through test, you continue with and complete the engineering review test following the guidelines provided here. When you complete your engineering review test, prepare the related

Grays

There are approximately 75 trillion cells in the human body, and each of these cells needs oxygen (among other elements) to function. When you receive intravenous chelation therapy--or even ingest oral chelating agents--you are helping all the cells utilize their oxygen. Nutritional scientists who administer the treatments do not know exactly how IV and oral chelators work. Only a few reactions are known relative to the vast numbers that exist, and the scientists are still studying them.

What is known is that since every body cell needs oxygen to function, by assisting in the oxygenation process, IV EDTA and oral chelating agents can, and do, significantly affect health and health-related problems, especially those involving the nerves, sex organs, and the cardiovascular system.

(*intravenous*)

4-7 *Sample document page after engineering review*

and necessary reports as discussed in the section of this chapter titled, "Tracking results and noting problems."

Usability test

Usability is defined as: 1. Capable of being used. 2. In a fit condition for use; intact or operative (American Heritage Dictionary).

A usability test is conducted to ensure that the documentation on which it is performed is capable of being used by its intended audience. Capable of being used means that it is easy to follow, and clear or easy to understand. Usability testing can, to some extent, verify that a document is usable.

The main drawback associated with usability testing is that it cannot reasonably be applied to entire documents, unless those documents are only a few pages long. For example, The Brinkmann Corporation sells an electric smoker, model number 810-5290-C, which I recently purchased (Brinkman 1989). The manual that accompanies the smoker is only 12 pages long, including the front and back covers. This document is limited enough in size to allow a usability test to be conducted on virtually every aspect of the document.

On the other hand, I also purchased a hot air circulating oven from a now-defunct company called Aroma Manufacturing Company. The manual that accompanied the AROMA™ AeroMatic Oven when I purchased it contains more than 50 pages (Chang 1991). To conduct a usability test on every page of this document would be very costly and time consuming. The larger the document, the more costly and time consuming the usability test will be.

This does not mean that you should not conduct usability tests on larger documents. In fact, the opposite might be true. The larger the document, the more important it is for the document to be easy to use. The usability test that you do conduct on large documents, however, will likely have to be very limited.

When conducting a usability test, there are six basic elements to keep in mind:

1 Development of problem statements or test objectives.
2 Use of a representative sample of typical users.
3 Representation of the actual work environment.
4 Observation and review of users participating in the test
5 Collection of performance and preference data.
6 Recommendation of design improvements. (Rubin 1994)

Based on these six elements, each usability test you conduct must start out with a clear problem or objective that is to be accomplished. For example, Fig. 4-8 shows a section of the "Brinkman Smoke 'n Grill Owners' Manual." Referencing this figure, you might choose to establish a testing goal that answers the concern of whether the user can successfully find information about basting foods while foods are being smoked.

The documentation tester generally plans, organizes, and conducts the usability test with typical users actually performing the tasks. This is different from any of the other types of documentation tests in that the documentation tester acts as a coordinator, but generally does not act as a participant. Sometimes the documentation tester is a participant. When this occurs, however, someone else in the company is usually responsible for planning, organizing, and conducting the test.

When the documentation tester is not a participant, but is instead the coordinator, one of the tester's first responsibilities is to prepare a list of tasks or assignments for the user to complete, using the manual.

The documentation tester, as usability testing coordinator, must also seek out and schedule "typical" users to conduct the usability

Cooking Tips

- Meats should be completely thawed before cooking.
- Marinating meats helps to break down the cellular structure, thus the cooking time generally is reduced.
- When cooking more than one piece of meat, the cooking time is determined by the largest single piece of meat being cooked.
- Brush poultry and naturally lean meats with cooking oil, butter or margarine before cooking.
- Save by cooking extra food for freezing. Properly wrap for later oven or micro-wave heating.

Flavoring Wood

To obtain your favorite smoke flavor, use chunks or sticks of flavor producing wook such as hickory, pecan, apple, cherry, or mesquite. Any fruit or nut tree wood may be used for flavoring. DO NOT USE resinous woods such as pine, for they will produce an unpleasant taste.

Seasoning

Use salt, herbs, and spices to your taste. You can get variations in flavor by adding wine, soft drinks, herbs and spices, bits of citrus peel or fruit juice, onion or marinades to the water pan.

Basting

The meat bastes itself during cooking on the water smoker. No basting, tending or turning is necessary after the meat is placed on the grid. For added variety, barbecue sauce or marinade may be used on the meat before placing it on the water smoker.

NOTE: For well done, cooking time may be longer. Add hot water to water pan as needed

8

4-8 *Sample page section from "Brinkman Smoke 'n Grill Owners' Manual"*

test. In addition, once the time comes to conduct the test, the testing environment should be set up so that it resembles the environment under which the product and documentation will be used. In the case of the smoker, the test might even be conducted outdoors, with smoker and documentation at hand.

Once the problem statements or goals are defined, typical users are chosen, and the environment is properly set up for conducting the test, the documentation tester's next responsibility is to observe the user during the test. For example, the test might require that the user be asked to accomplish a specific task using the manual, such as the following: Explain how to baste meat when using the smoker to cook instead of to smoke a piece of meat.

If the document is well designed, well written, and usable, the user should not have any trouble locating this information. If the user has any difficulty, goes astray, looks under topics different than what you expected, and so on, the documentation tester makes note of all these user reactions and attempts to find the needed information. The documentation tester even tracks emotional responses.

After conducting the usability test with a sufficient sampling of representative users, the tester is then responsible for collecting all of the resulting performance and preference data. He then makes recommendations related to improving the document's design so as to make it more usable.

Usability testing is a science and an art. The documentation tester usually participates in the usability test as a "typical" user, rather than as a coordinator. Usability testing is tremendously time consuming. When a documentation tester becomes involved in coordinating a usability test, there is usually little time left to conduct any other type of test. Therefore, usability testing is generally the last type of test a documentation tester will select.

Because of the time and effort involved in usability testing, several books and many articles have been written on the subject. This book cannot go into great detail about how to conduct a usability test. If you are interested in finding out more about usability testing, read one of the many available books. One particularly good one is Jeffrey Rubin's *Handbook of Usability Testing: How to Plan, Design, and Conduct Effective Tests* (Rubin 1994). You will also find many excellent sources of information written by Judith Ramey.

If you conduct a usability test, when you complete the test, you should prepare the related and necessary reports as discussed in the section of this chapter titled, "Tracking Results and Noting Problems."

Basic functionality test

The primary purpose of the basic functionality test is to ensure that the product functions as described in the accompanying documentation. The emphasis is on ensuring the accuracy of any task-oriented information the document provides, rather than on changing the product. Therefore, in order to conduct a basic functionality test, you must have a procedure or task-oriented document. While the entire document is not limited to describing and explaining tasks, and can include conceptual information, the basic functionality test can be applied to only the task-oriented information. The conceptual information will need a read-through or similar type of test performed on it.

As with all other types of tests, you must plan, prepare, and set up for the basic functionality test. The same planning, preparation, and setup that were discussed previously for read-through tests apply to conducting a basic functionality test as well.

Once you are set up and ready to begin the test, you must include one thing the read-through test does not always require—the actual product. You cannot test tasks and procedures to be performed using a product, if you do not have the product on which to perform those tasks or procedures.

Figure 4-9 shows a section of the "Garden Way Chipper/Vac Owner/ Operator Manual" for model number 47031 (Garden Way Inc. 1993). Based on Fig. 4-9, to conduct a basic functionality test of this section of the manual, you should look for and approach the test as described below and in the following pages, assuming the Garden Way Chipper/Vac is set up and functioning.

To Remove and Replace the Blade:

A. *Shut off engine, disconnect spark plug wire from spark plug, and make sure that all moving parts have come to a complete stop.*

B. Remove the collection bag or the blower deflector.

C. The blade removal procedure requires removal of the engine. To avoid a potential safety hazard from spilled gasoline, the fuel tank must be emptied of all gasoline before the engine is removed. Follow stpes C-1 through C-4 to empty the fuel tank.

1. Place a clean pan below the fuel tank to catch the gasoline as it drains. Remove gas cap from fuel tank.

 DANGER

Gasoline and its vapors are highly flammable and explosive.

To avoid serious personal injury, be sure area is well ventilated and keep gasoline away from open flame or sparks. Observe no smoking rules at all times. Use an approved fuel container.

4-9 *Sample page section from Garden Way® "Chipper/Vac Owner/ Operator Manual" before testing*

To conduct a basic functionality test on this or any procedural document, follow each of the steps or tasks exactly as described in the document. Using the information shown in Fig. 4-9, the first thing you do is shut off the engine. As a documentation tester, you should ask yourself whether the user needs to be told how to shut off the engine, or if the user will know that at this point in the documentation.

This section of the page is describing an advanced skill—removing and replacing the chipper/vac's blade. You should also know enough about the product to know that such a procedure only has to be done after the user has worn out or broken the blade from repeated use. You can assume, therefore, that the user knows how to shut off the chipper/vac by that time.

The next portion of the sentence, "disconnect spark plug wire from spark plug," is not a task the user might regularly perform. In this case, perhaps this part of the document should either explain how to remove the spark plug wire from the spark plug, as is done for instructions on emptying the fuel tank later on this page, or should reference the location within the manual of instructions for disconnecting the spark plug wire from the spark plug.

An experienced user might have changed the chipper/vac's spark plug on at least one occasion. If spark plugs are changed more frequently than blades, it might be safe to assume the user already knows how to disconnect the spark plug wire from the spark plug. If, however, you believe it best to provide a reference, then you would note that on the document.

The remainder of the first sentence instructs the user to "make sure that all moving parts have come to a complete stop." The only real moving part is the blade. And while you would not want anyone to try to remove the blade if it was still spinning, it might be just as dangerous for the user to reach under the chipper/vac's housing assembly to verify that the blade has stopped spinning. Therefore, it might be more technically correct (and safer for the user) to reword this portion of the sentence so that it reads similar to the following: *and wait until all moving parts have come to a complete stop before proceeding*.

The next section of this document instructs the user to "Remove the collection bag or the blower deflector." Again realizing that it is likely to be an experienced user who is changing the chipper/vac's blade, and that removing the collection bag or blower deflector is likely to have been done several times in the past by the user, no further reference to instructions for performing this task are necessary at this point in the document.

As a documentation tester performing a basic functionality test, you perform each of these tasks as instructed. Therefore, you shut off the engine, disconnect the spark plug wire from the spark plug, and wait until all moving parts have come to a complete stop.

You then attempt to remove the collection bag or the blower deflector. What happens if, during your attempt to remove the collection bag or blower deflector, you find that you cannot do so without first lowering the handle? As a tester, you would then add this instruction to the document, modifying the step letters as needed to include this information in the proper order.

Continue by performing additional instructions as presented. Although checking for spelling, grammar, punctuation, and so on is not your primary concern, if you come across a problem, you should point it out as well. For example, the word "steps" in the sentence that reads, "Follow steps C-1 through C-4 to empty the fuel tank." is misspelled. Note its correct spelling on the document.

Figure 4-10 shows the same section of documentation as shown in Fig. 4-9. In Fig. 4-10, however, all of the documentation tester's notes and comments, as discussed previously, have been included.

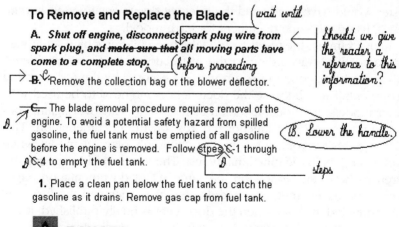

To Remove and Replace the Blade: *(wait until*

A. **Shut off engine, disconnect spark plug wire from spark plug, and ~~make sure that~~ all moving parts have come to a complete stop.** *(before proceeding* — *Should we give the reader a reference to this information?*

B. Remove the collection bag or the blower deflector.

C. The blade removal procedure requires removal of the engine. To avoid a potential safety hazard from spilled gasoline, the fuel tank must be emptied of all gasoline before the engine is removed. Follow steps C-1 through C-4 to empty the fuel tank. *(B. Lower the handle.)* *steps*

1. Place a clean pan below the fuel tank to catch the gasoline as it drains. Remove gas cap from fuel tank.

 DANGER

Gasoline and its vapors are highly flammable and explosive.

To avoid serious personal injury, be sure area is well ventilated and keep gasoline away from open flame or sparks. Observe no smoking rules at all times. Use an approved fuel container.

4-10 *Sample page section from Garden Way® "Chipper/Vac Owner/ Operator Manual" with tester's notes*

As with all other types of documentation tests, once you have completed your basic functionality test, prepare the related and necessary reports as discussed in the section of this chapter titled, "Tracking Results and Noting Problems."

Integration test

The last type of test is the integration test. This test is performed to ensure the quality of the documentation for different products or software platforms (operating systems), when the documentation has been designed and written to work with different products or different software platforms.

An integration test can include any or all of the other types of tests. The difference is that the test is performed for each of the different products or software platforms to which the documentation applies.

Remember the children's play set I mentioned in the introduction to this book. The instructions included with that play set were designed for several different models of the same product. If an integration test had been performed on that document, the documentation tester would have conducted the relevant documentation test several times, using the same document for each model.

For example, the tester would take one children's play set model, and assemble it according to the instructions. All errors, problems, missing information, and so on would be noted by the tester on the document. Then, because the document is intended to be used with other models, the documentation tester would start over again, conducting another basic functionality test on the documentation. This time, the tester would use another children's play set model when performing the basic functionality test. The tester would note any differences between models on the same set of documentation pages where the tester's notes and comments were included the first time he conducted the test. When the document is finally published, it will contain notes and comments specific to each individual model, as well as information common to all models.

If the documentation requiring an integration test is conceptual rather than procedural, then a read-through test should be used instead of a basic functionality test. In this case, the tester might only have to read the document once, providing he can account for all of the different products or platforms for which this document was written.

In some cases, an integration test is performed by several testers instead of just one. For example, assume a user's guide for a word processing software program was written to be used regardless of whether the software was running on a DOS-based computer or on a computer running the OS/2 operating system. If the tester is only familiar with DOS, the tester can conduct the read-through or basic functionality test using the software running on a DOS-based computer. Then, another tester can take the same document and conduct a read-through or basic functionality test using the software running on an OS/2-based computer.

Two tests of the same type conducted on a document to verify its accuracy against different products or different platforms is another definition of an integration test. Both testers perform the same tests, but with varying backgrounds, skill and knowledge levels, and products or platforms.

One thing to keep in mind when conducting an integration test is that you might want to make your comments and notes using a different color ink than that used by another tester or for a different product or platform. This approach helps both the tester and the writer to sort out different comments and notes if any confusion should exist. It also helps reduce the number of retests of sections you might have to do in order to figure out which comment related to which product or platform.

As with all documentation tests, when you complete the integration test, you must prepare the related and necessary reports. If you conduct all of the integration test, you can prepare one report that reflects comments, changes, suggestions, and so on associated with all products or platforms. If more than one tester is needed to conduct a complete integration test, then each tester generally writes their own testing report, although they might work together to produce a single, combined report.

Tracking results and noting problems

Completing a documentation test can be fun, informative, interesting, and useful to all concerned. A completed documentation test is only useful, however, if the results of that test are reported and acted on.

When reporting the results of a documentation test, there are three key issues that should concern you:

- How you will record errors you found in the document
- How you will note areas of the document needing improvement
- How you will report any problems you find that are directly related to the product

Recording errors in the documentation

When testing a document, regardless of the type of test you are conducting, be consistent in how you note errors, changes, corrections, and so on. If your documentation test does not require that you be very formal about making your notations, you can draw attention to those changes using such simple techniques as lines with arrows, circles, comments in the margin, and so on. If you must use a more formal method of marking the documents, one method of doing so is to use proofreader's marks.

Proofreader's marks are standardized symbols for indicating changes that must be made to a document. You are probably already familiar with some of these marks. For example, lining through a word or words, then placing what looks like a lowercase e with tails on both ends in the documents margin next to the sentence containing the lined-through words is a standard method (Fig. 4-11) for indicating that text is to be removed from a sentence.

You can find information about proofreader's marks in dictionaries and writing reference manuals and books. In the *American Heritage Dictionary*, a table of proofreader's marks is located on the same page where the word *proofread* is defined (American Heritage Dictionary).

Noting documentation areas that need improvement

When you conduct documentation tests, you will come across areas in the document that need to be improved. They might be missing a substantial amount of information, or information which, while not lengthy, is important to the document. When this occurs, you have two options.

The first option is for you to simply make suggestions as to the best way to improve the document. For example, if you are testing a document and find that key words are hard to locate, you might make a note of the problem to the writer. Then, you might make a suggestion or two about how to correct the problem. In the case of key words being difficult to find, you might suggest that the author set these words in a bold typeface, or that he underline them so they are easier to locate.

Some Common Proofreader's Marks

Purpose	Mark in Margin	Mark in text	Example
Delete text	⌒	——	A ~~bad~~ dog
Capitalize	(cap)	≡	A bad dog
Make lowercase	(lc)	/	A Bad dog
Do not delete	(stet)	....	A bad dog
Insert a period	⊙	∧	A bad dog

4-11 *Common proofreader's marks*

This is the method most often used. In the document you indicate that some changes or improvements need to be made, and to see the report for more details. Then, the report would explain what the problem is, and provide a suggestion or two for correcting it. If possible, the report should also include a sample that shows one way your suggestion might be implemented.

The second option is to make the actual correction yourself. Using the same example of key words being difficult to locate, you would mark every key word to be boldface. Using standard proofreader's marks, you would place a wavy line under the key word, then place the letters b and f inside a circle in the margin to the left of the text.

Noting product problems

When you test a product's documentation, you are likely to find problems with the product, in addition to finding problems with the documentation. There are several methods for reporting product problems. Some of the most common methods include

- Talking directly with the responsible engineer or designer
- Formally presenting the problem in a development meeting
- Preparing a written description of product problems, including any suggestions you have for correcting the problems

- Entering information about the problems into an electronic database, if one is available, or filling out a form designed for this purpose

The reporting method you choose depends on company policy. It also depends on the availability of reporting alternatives. For example, a company requiring a formal notice of the problem might also have an electronic database into which you enter information about any problems you find. On the other hand, formal notice might be presented by filling out and delivering to the proper party a form the company has created.

If your company has no formal method for reporting product problems that are found during the documentation test, try reporting the problems through a more informal channel. Regardless of the approach you take, report product problems you find. Do not simply assume that someone else will find and report them.

The report you prepare to accompany the documentation you tested should include information related to both the good aspects of the document, as well as the bad. Start your report by explaining what you found to be particularly good, interesting, or effective about the document you were testing. Establish a positive attitude with your report.

After discussing what is good about the document, you can then discuss the problems you found in the document. The report is not the place to list each page and each problem, it is the place to provide information about problems that are general to the document as a whole.

For example, you might include a section in your report titled "General Concerns," or something similar. Under this heading, provide information relevant to the document as a whole. This is a good place to present your belief that key words in the document should be bold or underlined, if this is a concern you have with the document.

If the document you tested is small in size or scope, you might include a bullet list of the specific pages on which you found problems. This bullet list, along with the flags you placed on the changed pages, makes it easy for the writer to find and modify the areas of the document on which you made notes or changes. The writer can use the bullet list as a checklist to ensure that all of the changes and comments you made have been found and incorporated.

How you present the report does not really matter. Of course, a formal written document is usually better than simply explaining the corrections or changes to the writer. However, it should not be used alone. It is always best to arrange a time to meet with the document's writer and review the document. Explain why you took a certain approach, make sure he can read your handwriting, review anything he has questions about, and so on. This makes it easier for the writer to implement your changes, corrections, and suggestions. It also makes it easier for the writer to accept the results of your test. Knowing that you believe in his abilities as a writer, and that your comments, changes, and so on are simply intended to improve the document and not to criticize his work, makes your relationship with the writer a more pleasant one. It also helps to ensure that the writer is open-minded and willing when it comes to implementing your changes and suggestions.

Summary

The purpose of conducting a documentation test is to ensure the quality of the document. As a side benefit, documentation testing often contributes to the quality of the product. When conducting your documentation test, always keep in mind the four basic items that affect a document's quality—technical accuracy, completeness of information, usability, and document clarity.

You should keep these four basic items in mind not only when you conduct the test, but when you choose the most appropriate test to conduct. Whether you choose to conduct a read-through, engineering review, usability, basic functionality, or integration test, choose the one that is most suitable for the type of document you are testing—conceptual, procedural, or quick reference.

Prepare for the chosen test by learning everything you can about the product, setting up an area in which the test can be conducted, writing a documentation testing plan, gathering anything you need to conduct the test, and so on. Then conduct the test you planned, and report the results both on the document and in the testing report. If, in the process of conducting the documentation test, you find problems with the product, be certain to report these as well. You cannot assume that someone else will do so.

By understanding how to conduct each type of documentation test, and by thoroughly preparing to conduct the chosen test, you can add substantially to the quality of the product documentation. Your role as a documentation tester is to ensure the quality of the documentation. In the process, you can help to ensure the quality of the product as well. In the end, both the product and the documentation will be better for the effort and attention you give to them.

References

American Heritage Dictionary, 2nd college edition, s.v. "proofread."

American Heritage Dictionary, 2nd college edition, s.v. "usable."

Brinkmann Corporation. 1989. "Brinkman Smoke 'n Grill Owner's Manual." Dallas, Texas: The Brinkmann Corporation.

Chang, Peter C.Y., ed. 1991. *Cook the Turbo Way*. Anaheim, California: Aroma Manufacturing Company.

Garden Way Inc. 1993. "Chipper/Vac Owner/Operator Manual." New York: Garden Way Inc.

Rubin, Jeffrey. 1994. *Handbook of Usability Testing: How to Plan, Design, and Conduct Effective Tests*. New York: John Wiley & Sons.

Trowbridge, John Parks, and Morton Walker. 1991. *The Healing Powers of Chelation Therapy: Unclog Your Arteries, an Alternative to Bypass Surgery*. Stamford, Connecticut: New Way of Life, Inc.

Williams, Peter L. and Roger Warwick, eds. 1980. *Gray's Anatomy*. 36th ed. Philadelphia: W. B. Saunders.

5

Reporting and following up on your testing results

No matter how wonderful your documentation test is, if the changes, suggestions, corrections, and so on that you make are not implemented, the test is almost useless. The most important part of the testing then, might be yet to come. You must now report what you found during the test, and get the writer and other responsible persons to accept your testing results and implement your changes and suggestions. One of the best ways to do this is by producing a useful and effective documentation testing report.

As noted in Chapter 4 there are three key issues with which you should be concerned when you conduct your test and produce the documentation testing report:
- How you record errors you find in the document
- How you note areas of the document that need improvement
- How you report any product problems you find

The report you prepare to accompany the tested document should include information about the errors you found in the document, as well as about areas of the document that need to be changed or improved. In addition, the report should contain references to any product problems you found while you were conducting the test.

The report should explain errors, improvements, and product problems. Part of the design of a report can include whether you are even going to prepare a written report. In some instances, it might be better to present your report verbally, as an accompaniment to the tested document. However, in my 20-plus years in this field, I have always found it best to provide a written report of the test results, even if that report is only a single page in length.

This chapter discusses documentation testing reports, including what should go into them, how they should be presented, and when verbal reports are acceptable. After reading this chapter, you will understand

- When to give verbal instead of written reports
- How to construct a documentation testing report
- How to report your testing results without making enemies
- How to report product-related problems
- How to follow up after the test

Giving verbal versus written reports

As stated previously, you should always provide written reports that explain the results of your documentation test. There is one instance, however, when a verbal report is appropriate. Presenting the writer with a verbal report instead of a written report is appropriate when you have conducted a preliminary test on the document or a portion of the document.

For example, some documents are so large or the project so lengthy that in order to keep everything on schedule you must pretest as much of the document as you can. This generally means you test the document a chapter or a section at a time. When this occurs, it is perfectly acceptable to return the tested chapter with a verbal instead of a written report.

Verbal reports

When you test in sections and provide verbal reports, you should still make notes for yourself regarding what you told the writer. When the writer makes the changes or corrections you suggest, and you receive the final document for its full documentation test, you will have your notes to reference to make certain any problems you noted have been satisfactorily corrected. Keeping some type of notes regarding the results of your pretests also limits how extensive the final test must be, reducing the amount of time it takes you to complete the final test.

Also note that these section tests should be considered preliminary tests. Some writers will ask you to do a final test on each chapter or section as it is presented to you, rather than waiting until you
receive the entire document. The writer who makes such a request is
generally doing so to save time, not to try to sneak something past
you. However, you cannot provide a complete test until you have all
the pieces of the document. Without the entire document in hand,
you will be unable to check cross-references or index entries, or to
match section titles with those shown in the table of contents.

The biggest problem with trying to conduct a thorough test on an
incomplete document is that it is almost impossible for you to tell if
something is missing. Although you can try, you cannot always correctly second-guess the writer. In order to ensure that a document
contains everything it needs to contain, you need to have the entire
document in hand. Therefore, you must insist on testing the entire
document when the writer has finished it.

When providing the writer with a verbal report, present the same
types of information you present if you are providing a written report.
Most of the time, this means you should include the following types
of information:

- A summary of what is good about the tested section or
  chapter
- Information about general problems discovered while testing
  the section or chapter
- Specific pages or items in the section or chapter that need to
  be changed or corrected, and why

Although you are verbally presenting the writer with details related to the section or chapter you tested, this is not the final test you
must give it or the final report you must present. The final test you
conduct must be conducted on the entire document, although the final test can then be more limited than it might otherwise have been.
After you complete the final test, it is time to provide the final test report, and to provide it in writing.

Written reports

Documentation testing reports can be formal or informal. They can
contain a wide variety of information, or they can contain a limited
amount of information. Written reports can outline the results of the
test (Fig. 5-1), or they can provide a page-by-page test review.

The best method for presenting your written findings depends on
what your company prefers. It also depends on how much informa-

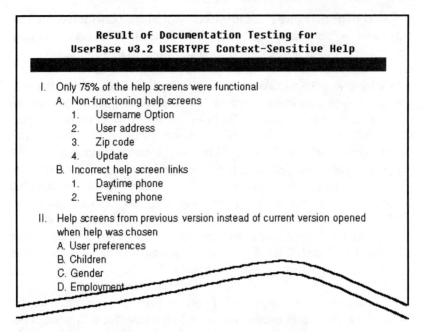

5-1 *Sample outline style documentation testing report*

tion you need to present to the writer, and whether or not the document will accompany the test report.

In most cases, only the writer is concerned about the results of the test. As the writer is also the one who must make the changes and corrections, the original document with your test changes and corrections noted should be returned to the writer along with the testing report. Therefore, the test report can be more of a summary than a detailed review of the document.

If the test report must go to someone in addition to the writer, it might need to provide more detail. For example, if the test report is to go to the writer's manager, you might need to be more detailed about your explanations. On the whole, however, most test reports can be one or two pages in length and provide general information. The document itself reflects the details as you noted them during the test.

Sample written report

A formal written report, regardless of its length, should include several basic items of information. As a document, the documentation testing report falls into the general category of reports that provide analysis information. In addition to the formal documentation testing report, other types of formal analysis reports include

- Project reports
- End-of-job reports
- Status reports
- Product design reports
- Project analysis reports

Although of different titles and different content, these analysis reports can all be presented using a similar basic format. The format for a formal analysis report should contain the following (Sides 1991):

- Title page
- Abstract
- Table of contents
- List of figures (if applicable)
- Introduction or overview
- Discussion of the topic
- Conclusions
- Recommendations
- Appendix

Title page
The title page should clearly identify the report attached to it (Fig. 5-2). "It should provide enough information so that readers can tell what the context of the report is and what the report is about" (Sides 1991). It should be designed so as to be pleasant to look at, but not distracting.

Abstract
The abstract should provide a succinct report of what was determined, how it was determined, and the significance of the findings. It is written to provide enough information to the reader so that the reader can determine whether or not this particular report is appropriate for them to read (Fig. 5-3).

Table of contents
The table of contents provides the reader with a quick glance at the main topics and subtopics covered in the report. It also makes it possible for the reader to find what he needs in the report, without having to read it from beginning to end.

List of figures
If the report includes figures, tables, special symbols, and so on, this section of the report lists those items. It should include an identifying number, a title or description, and a page reference.

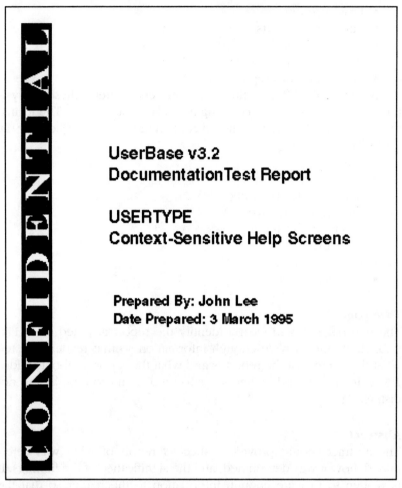

5-2 *Sample documentation testing report title page*

Introduction or overview
This section tells the reader what information within the report is important. It also orients the reader to the report's contents, and lets him know what information he can expect to get from reading the report.

Discussion of the topic
The discussion of the topic is the most important part of the report. It contains all relevant and important information regarding the report's topic. In a documentation testing report, this section should include sections from the document that are used to support statements made relevant to the document.

ABSTRACT

UserBase v3.2 is an application program providing database functions and features to small businesses and home computer users. An important feature of UserBase v3.2 is its online context-sensitive help. Each UserBase v3.2 menu provides access to context-sensitive help.

This documentation test report provides the results of the documentation test conducted on the USERTYPE menu for the UserBase v3.2 program. Overall, the USERTYPE help screens are not sufficiently complete to ship with this product in their present state. Approximately 25% of the online help screens are unfinished or incorrect. This problem must be corrected before the USERTYPE menu can be shipped with this product.

5-3 *Sample abstract for documentation testing report*

Conclusions

This section states the results of the analysis. In the case of a documentation test, this section reports the results of the test in a summary form. Specific issues can be listed and then briefly discussed (Fig. 5-4).

Recommendations

If appropriate, this section explains what should be done based on the findings of the analysis. In the case of documentation testing, this section can provide a list of methods the writer could use to improve the quality and accuracy of the document.

Appendix

Information that is not vital to the report, but which might be useful to the reader of the report can be placed into an appendix. When preparing a documentation testing report, an appropriate appendix might include one or more sample pages from a competitor's documentation.

Conclusions

The context-sensitive help screens for the USERTYPE menu are incomplete (only 75% are finished). In addition, those help screens which are complete should be modified as follows:

All screens which contain the statement "To cancel, press F7." need to be modified to read "To exit, press F7."

Once the additional screens are prepared, they will need to be tested. In addition, all changed screens will need to be retested.

Some consideration should also be given to ways in which the help screens can be

5-4 *Sample conclusions section from a documentation testing report*

Constructing the report

When you prepare your documentation testing report, there are several factors to keep in mind. It also helps to know what a final report might look like. The figures shown throughout this chapter provide you with samples of sections from a formal analysis-type report. Documentation testing reports, although they fall into the analysis category of reports, might include some information not listed as part of that type of report. You must decide what information is most appropriate for the testing report you are preparing.

When constructing your report, you should consider the following suggestions:

- Know the audience
- Provide a test overview
- Summarize the results
- List good points
- List and reference specific problems
- Report product problems
- Make recommendations
- Summarize the report

Know the audience

When preparing the documentation test report, you need to keep in mind the intended audience. Unlike the document you are testing, the intended audience is not the product's user. Instead, the intended audience is company personnel.

Just as it is important when preparing the documentation test plan to know the document's intended audience, it is also important to keep in mind the test report's intended audience. Most commonly, that audience is the tested document's writer. However, there might be several individuals who receive, read, and review the results of the documentation test. That audience can range from members of the development and documentation teams to the company's executive staff. More often than not, however, the test results are reported only to the document's writer, so your primary audience is usually the writer.

Although you need to remember who the audience is for this testing report, you do not need to specify that audience anywhere in the report, with one exception. If you need to include a list of people to receive the report on the cover sheet, then you need to specify the audience. Otherwise, the report does not include a list of potential readers.

Provide a test overview

The report should include a description of the test. The information contained in this section can be brief, explaining only the document that was tested and the platform on which it was tested, or it can be rather lengthy. The following is an example of a brief test overview:

After reviewing the USERTYPE context-sensitive help screens for the DOS-based version of UserBase v3.2, the following issues and concerns need to be addressed.

A brief test overview is appropriate when the scope of the test is limited. For example, if the test only covers a single aspect of the product, as in the previous example, even a single-sentence overview is sufficient.

Writing a brief overview of the test is also appropriate when those who will be reading the test results are already very familiar with the test you conducted. For example, documentation team members, including the document's writer, should have reviewed the documentation test plan you prepared. This plan should have given every reader a detailed overview of the test to be conducted. Under these circumstances, a lengthy overview of the test you conducted is not necessary.

If the documentation test report is to be distributed to individuals who did not have an opportunity to read and review the documentation test plan, then you should consider lengthening the test overview in the report. In this case, additional information, such as the audience definition, the product's competition, the level of test that was conducted, and other relevant information, should be included in the test overview of the report (Fig. 5-5).

- 2 -

Test Overview

The UserBase v3.2 USERTYPE context-sensitive help screens were tested to determine the technical accuracy, usability, completeness, and clarity of the information they contain. The testing was conducted based on the audience for which these screens were written, the competition this product has, and the level of test deemed appropriate.

Audience

The UserBase v3.2 application program is a database program designed for small business and home users. It is also designed to be used by small groups or teams of people within larger companies.

Competition

UserBase v3.2 has as its chief competitor dataB v2.x. This product is being marketed to the same primary and secondary audiences as is UserBase v3.2.

Level of Test

A basic function
to dete

5-5 *Detailed test overview in a documentation testing report*

Whenever possible, keep the overview to one or two sentences. If you need to provide additional information in the overview, consider attaching a copy of the original documentation test plan to the test report for those individuals who need the additional information. This approach minimizes the relevant but less important information in the test report, but provides additional information for those who need it.

Summarize the results

This section of the report should include a summary of what you found when testing the document. The size of the summary depends on several factors:

- The size of the overall report
- The number of problems found in the document
- The variety of problems found in the document

If you are preparing a single-page documentation testing report, you are obviously limited as to the size of the summary it includes. In such a small report, the summary might consist of only a single paragraph or even just a sentence or two.

If the report is slightly larger, two to five pages for example, the summary can be larger (Fig. 5-6). A paragraph or two followed by a bullet list of major problems might be appropriate.

- 5 -

Summary

Overall, the help screens for the UserBase v3.2 USERTYPE utility are easy to read and quick to access. In addition, the screens that could be tested are complete and well written. There were some problems; however.

After completing a basic functionality test on the UserBase v3.2 USERTYPE context-sensitive help screens, the following issues became evident:

- Only 75% of the help screens are fully functional. Of the 25% not fully functional, the majority do not open at all. An error message indicating "No help currently available" opens instead. The balance of the non-functional screens have screens which open, but which are not the correct screen.

- Four of the screens were help screens from the previous version of this product.

- All instances of the message "To changed to instead

5-6 *Sample documentation testing report detailed summary*

In a large report where every item found during the test is discussed, the summary can be as large as a single page. However, there is a point at which the summary is no longer a summary. If you need to create a summary larger than a single page, you should consider whether the summary contains too much detail.

List good points

Everyone likes to know what they did well. All documents have something about them that is good or was done well. It helps the writer to be more open to the negative comments you have about their document if you start with positive comments. Listing what you think is especially effective or well done in the document is a good idea. Not only does it help the writer feel better about the test in general, but it also gives you a positive topic to open your conversation with the writer when you go over the results of the documentation test.

List and reference specific problems

Now that you have provided a positive note on which to start the report, you can begin listing and describing the problems that you found during the test. At this point in the report, you should be listing and describing only those problems that occur in the documentation. This section of the report should not include information about product problems, with one exception.

If the document is incorrect because it reflects the product as it will be, instead of as it is on the day of the test, then you should point out the discrepancy between the product and the documentation. As a development team member, you should be aware of any major changes the product is currently undergoing. You should then be able to tell if the writer has chosen to document the planned result of these changes, instead of the current status of the product.

For example, if the product is being shipped in only one color, and that color is currently changing from green to gold, the writer might choose to refer to the product's gold color in the documentation. Because the product you are testing the documentation against might not yet be produced in the gold color, the documentation is incorrect. However, because you and the writer know the color is being changed, you simply remind the writer that if the product color change is not implemented, this section of the document will have to be changed at a later date.

Although making such a note might seem pointless, consider the consequences of not making such a note either on the documentation or in the report. Assume that for some reason the manufacturer cannot get the gold coloring it needs to produce the product on schedule. The company might decide to go ahead and ship the first several hundred products in the original green color. Soon after the product is on the shelf, your company's service and support lines start getting calls referring to the gold color of the product. Callers want to know if something is missing from the box, or how they can go about trading in the green product for a gold one. The number of calls escalates, and soon management becomes involved. Management wants to know who sent out this document with such an obvious error in it, because it is costing them thousands of dollars in service and support calls. Are you starting to get the picture?

Mud flows downhill, and you do not want to be standing at the bottom when it hits. If your report and the document indicate that you noted the potential discrepancy and brought it to the attention of those concerned, you might not need (or receive) a bath. If it appears as though you overlooked, missed, or ignored the problem, however, you can expect there to be some unhappy people, one of whom is likely to be you.

Okay, yes, it is a cover-your-tush exercise. But the reality is, you have to do it, like it or not, to keep it from coming back to haunt you in the future. Therefore, if you find discrepancies between the documentation and the product that are intended to be resolved by the time the documentation is complete, do not necessarily mark them as problems, but do note that you noted the discrepancy.

Report product problems

If the problem you found is a problem with the product as opposed to being a problem with the documentation, then you need to report this in the documentation testing report. You report product problems not only as a method of protecting yourself, but primarily as a way of contributing to the quality of the product.

Noting the product problems you found as part of documentation testing makes the writer aware that you found product problems, and that you will be reporting them to the proper person or through the correct channels. It also helps the writer to understand potential product problems affecting the documentation. In addition, it might also help him understand why you noted problems or discrepancies in his document when he had not seen the related product problem himself while writing the document.

Make recommendations

For the document that you test, you make comments, suggestions, and recommendations for changing and improving the document. Most of these recommendations are specific to the particular area or portion of the document you are testing at the time. The documentation testing report is a good place to put any general recommendations you have.

The documentation testing report is also a good place to put any specific recommendations you have that effect the document in many places, instead of just a few. For example, Fig. 5-4 shows a general recommendation made as the result of a documentation test conducted on the help screens for the USERTYPE software menu. Instead of marking every location in the document where this problem and its recommended solution occurred, the tester noted the first few. Once the pattern and consistency of the problem became known, the tester discontinued noting it at each place in the document, and instead made it a recommendation in the documentation testing report. The writer can then implement the correction by using the search-and-replace feature of his word processing software. This approach ensures that all occurrences in the document are found and corrected, corrects them all exactly the same way, and reduces the amount of time it takes both the tester and the writer to find and fix a problem.

Summarize the report

The last section of the documentation testing report should be a brief summary of the results of the test. It should quickly review the points you consider important. Most of the time, a single paragraph is sufficient.

The purpose of summarizing the report is to once more point out the important aspects of the testing results, as well as to give a feeling of being finished to the report and to the people who read it.

Presenting findings without making enemies

Once you finish the documentation test and prepare the testing report, your next responsibility is to see that the writer understands what your report and testing results indicate. The best way to ensure the writer understands the changes, corrections, and recommendations that are in the document and its accompanying report is to present it to the writer in person, and go over it with the writer. The presentation of your test findings is in and of itself a process (Fig. 5-7).

Meet with the writer

The first step in presenting your test findings is to set up an appointment to meet with the writer. Call, e-mail, or visit the writer. Explain that your test is complete and that you want to go over it to ensure that everything is legible and understandable. Try to schedule the appointment around the writer's schedule. You are likely to have your

Presenting Documentation Test Findings

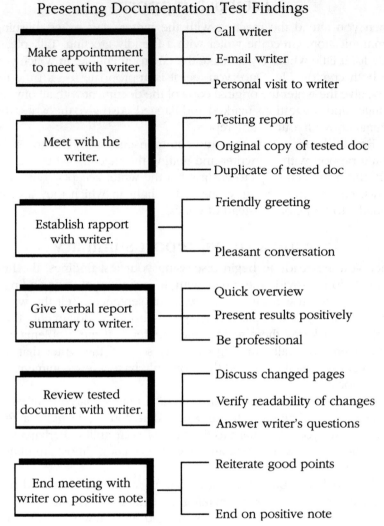

Make appointmment to meet with writer.	Call writer E-mail writer Personal visit to writer
Meet with the writer.	Testing report Original copy of tested doc Duplicate of tested doc
Establish rapport with writer.	Friendly greeting Pleasant conversation
Give verbal report summary to writer.	Quick overview Present results positively Be professional
Review tested document with writer.	Discuss changed pages Verify readability of changes Answer writer's questions
End meeting with writer on positive note.	Reiterate good points End on positive note

5-7 *Process of reporting test findings*

changes and corrections accepted in a more positive light if you are as accommodating to the writer as possible.

Once you have scheduled the appointment, prepare to attend the meeting. The most important preparation for you to make is that of copying your testing report and the document you tested. Leave the original of the report and the tested document with the writer, and keep a copy for yourself. You will need these copies to use during the meeting with the writer, as well as later on when it comes time to check the changes the writer made to the document based on your test.

Establish rapport

When you attend the meeting with the writer, start by establishing communication. Greet the writer with a friendly greeting. Talk pleasantly for a little while. Try to make the environment one in which you are both comfortable. Once you feel it is appropriate to start the review, give the writer the original copy of the document with all of your changes and suggestions marked and flagged. Also give the writer the original copy of your testing report.

Besides being a courteous and commonsense approach to establishing rapport with the writer and getting the meeting off to a good start, this approach helps to ensure that the writer accepts your comments, changes, and suggestions in the light in which they are intended—to be helpful instead of critical.

Provide a verbal report summary

Once you are ready to begin discussing your test findings, the first thing to do is provide the writer with an overview of your findings. Summarize the test results in a few quick sentences. Tell the writer what you thought was particularly good about the document. Then let the writer know there were some problems. If the problems you found were not really any big deal, say so. Tell the writer that the document was well done and that you only have a few comments or suggestions.

If, on the other hand, the document contains a large number of corrections or changes, tell the writer that you made quite a few notes. Buffer your statement to the best of your ability. Perhaps you can indicate that, while there are several corrections, most are minor, if this is the case.

The goal is to present the information in a positive instead of a negative way. The reason for making the extra effort at being positive when you present your findings is twofold. First, few writers are required to incorporate your changes. If they do not agree with your comments or changes, they can choose to ignore them.

Second, there is no need to make this any more painful for you or the writer than it has to be. Make every effort to turn the meeting into a positive and pleasant experience. This approach does not lessen anyone's opinion of your abilities as a documentation tester. In fact, the result is generally just the opposite. The writer is more likely to think of you as a professional, doing your job to the best of your ability, concerned with what is best for the product and the documentation, and not just with making yourself look good.

Review the document with the writer

The next step is to go over the document with the writer, page by page, skipping any pages on which you have not made any changes or notes. (This is another good reason to flag changed pages. You do not have to thumb through each page of the document trying to find your changes. Instead, you can open directly to them.)

When reviewing the document with the writer, the first goal is to make certain the writer can read your changes and comments. The second goal is to explain the reasons why you made any changes or suggestions that are not self-explanatory.

As a rule, you do not want this meeting to last for hours. One good way to shorten it is to give the entire document to the writer. Explain that the changed pages have flags on them, and that you will answer any questions or provide any input related to any of the suggestions you made. You can then sit back and let the writer review the flagged pages at a comfortable pace. If you take this approach, however, watch the writer. If the writer gives you the impression that he is uncomfortable with anything he is reading, chime in with an explanation of why you made a particular suggestion or change at that point in the document. Of course, this means you have to be following along, changing pages in your copy of the document. The goal is to help the writer to know that none of your changes or suggestions are meant to be taken personally, and that any changes or corrections that are not appropriate can be eliminated.

When you meet with the writer, be prepared to have the writer refute some of your changes or comments. If the writer can explain why something was done the way it was, and the explanation leads you to believe that the change you suggested should not be made, make a note to yourself about it on your copy of the tested document. Let the writer know you agree with the decision not to make the change or implement the suggestion. No one expects you to be perfect. To be honest, it sometimes helps the writer accept that changes need to be made to the document when the writer sees that you too make an occasional mistake.

End on a positive note

When you and the writer have finished reviewing the changed document and the report, try to end the meeting on a positive note. Unless your career as a documentation tester or your time with this particular company is short, you are likely to be working with this writer again some time in the near future. You do not want the writer to despair when he hears you have been assigned as the tester on his next document. Instead, you want the writer to prefer you to test his documents so that he knows they are the best they can be.

Just as you started the review by telling the writer what you thought was particularly good about the document, you can end the review by reiterating how well the document was written, designed, presented, or whatever was particularly good about it. The goal is to end the meeting with a comfort level that makes the writer anxious to input your corrections, changes, and suggestions.

Reporting product problems

It is rare that you test a product's documentation and never find a single problem or potential problem with the product. When you do find a problem, the problem needs to be reported. You report the problem so that the opportunity to correct it is made available. You also report the problem so that no one questions you later as to why you did not report the problem.

There are several methods for reporting product problems. As Chapter 4 pointed out, some of the most common methods of reporting product problems include

- Talking directly with the responsible engineer or designer
- Formally presenting the problem in a development meeting
- Preparing a written description of product problems, including any suggestions you have for correcting the problems
- Entering information about the problems into an electronic database, if one is available, or filling out a form designed for this purpose

The reporting method you choose depends on company policy. It also depends on the availability of reporting alternatives. Whether you choose to report product problems formally or informally is affected by whether your company has established a formal method for reporting product problems. If a formal process is defined, follow that process. If your company does not have a formal process for reporting product problems or potential problems, you should attempt to report the problems through a more informal channel.

Regardless of the approach you take, report any product problems you find and document, for your own records, that you found and reported the problem. If you do not report the problem, but instead simply assume that someone else will find and report it, be prepared to explain why you did not report the problem when you noticed it during your documentation test.

To or through the writer

In addition to reporting problems related to the documentation, you can report product-related problems to the document's writer. The writer is then responsible for reporting the problem through the appropriate channels. This ensures that the problem is brought to someone's attention.

Whenever possible, attempt to report product-related problems to someone in a position to verify and correct them. If you do not know who that person is, or how to contact such a person, reporting the problem to the writer is the next best thing to do.

Directly to the engineer

Your best alternative might be to report the problem directly to the engineer responsible for the product. If you are regularly attending product development meetings, you should know who is responsible for this aspect of the product. Report the potential product problem to that individual.

If that individual is not available for you to speak with, send the engineer e-mail or a memo. Put the information related to the potential problem in writing whenever possible.

Your information about a product problem probably will not be received with a warm reception. It might even be ignored. If this happens, you can choose to escalate the reporting to the next level—telling the development team.

In development meetings

It is sometimes easier to turn your back on a potential problem than it is to follow up on it and make sure it receives appropriate attention. Avoid this easy way out, particularly if the problem effects the quality of the product or the user's perception of the quality of the product. This means you might have to bring the problem to the attention of the development team members. If you do, you can present the problem verbally instead of in writing at a development meeting. Not only will all team members be made aware of the problem, but the meeting minutes will reflect that you made an attempt to point out the problem, rather than just ignoring or burying the information.

Through an in-house software bug database

Some companies have a computerized method for reporting product problems. In companies that produce software, this type of database is common. It is often referred to as a software bug database.

Regardless of the type of product your company creates, if a software (electronic) version of a database for reporting product problems exists, find out how to use it. Get permission to use it, then use it to report and track any product problems you find while conducting your documentation tests.

Through e-mail

Many companies now have some form of e-mail. If your company takes advantage of this method of interoffice communication, you should as well. You can send an electronic letter or short note to the appropriate party, whenever you find a product problem (Fig. 5-8). Keep a copy for yourself (or include yourself on the list of electronic recipients so that a copy is delivered to your electronic mailbox) for future reference.

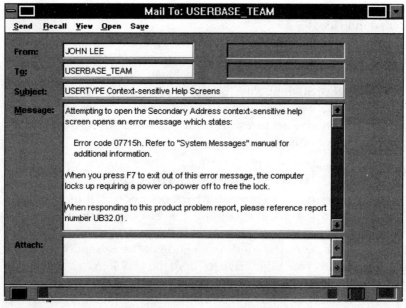

5-8 *Sample e-mail product problem report*

You have several options for reporting product problems. Choose the one most compatible with your company and its policies, then use it to report any potential product problems that you find.

Following up after testing

Once you complete the documentation test, prepare the report, and present your findings, you can sigh with relief; but you are not quite finished. You still must follow up after the test to see that your changes and suggestions are incorporated into the final document. If they are not incorporated, you need to make a note of it for the next release of the document, or at the very least, ask for an explanation of why the changes or suggestions were not included.

What to follow up on

You should follow up on every change and suggestion you made to ensure that nothing was missed or misinterpreted. As far as the document is concerned, this means you should check every change that was made, as well as those that were not.

You should also follow up on product problems you reported. Even if the engineers decide not to correct any problems you found, you should know that such a decision was made. It is also important that you know why they made the decision not to correct a problem.

If it is simply not possible to make the correction, there is no point in mentioning the problem again. If, however, they simply could not make the correction for this release of the product, your diligence in following up can help to ensure the problem is corrected for the next release.

Check the changes

When the author has finished making changes to the document, ask for a new copy of it. You should also ask to have the test copy returned to you. Check each change the writer made. Be sure the writer did not misinterpret or misread any corrections or suggestions you made. Sometimes, because changes and corrections are not properly handled, the change can make the problem worse instead of better.

Conduct regression tests

A *regression test* is a retesting of the document, or of certain parts of the document. Bill Hetzel defines regression testing as, "Selective testing to verify that modifications have not caused unintended adverse side effects or to verify that a modified system still meets requirements" (Hetzel 1988).

This time when you test sections of the document, you test only the sections of the document that were changed, and then only to verify that the change is correct. You can regression test any area of the document you believe needs to be retested. You do not have to regression test simple changes with limited impact. However, if you suggested the writer make a specific change that has a significant impact on the document, you should consider retesting the areas of the document effected by the change.

Regression tests are also conducted for the sake of checking a product change, rather than a document change. If you found and reported a potential product problem that was then corrected, you should retest the document against the product to ensure that both are now accurate and adequate.

Prepare the final report

Once you review the document changes, you should prepare a final report. This final report might simply state that all changes as noted in the tested document were correctly made. Most likely, however, there will be some changes that cannot be made, or which can be made but not in time for the release of the document. If this is the case, this information should go into the final report. This report can then be used to make sure everything is done to promote the quality and accuracy of future releases of the document.

Summary

This chapter discussed documentation testing reports—what should go into them and how they should be presented—because no matter how wonderful your documentation test is, if your changes, suggestions, and corrections are not implemented, the test is almost useless. One way to help ensure the implementation of your changes and suggestions is to prepare and deliver a succinct and effective documentation testing report.

To help you understand the importance of the report, as well as to provide suggestions and information on how to prepare an effective report, this chapter discussed various aspects of documentation testing reports. It explained when it was acceptable to give a verbal instead of a written report. It also showed you an example of how to construct a documentation testing report. In addition, it provided information to help you understand how to report your testing results without making enemies. As a final note on documentation testing reports, it explained how to follow up after you complete the test, and the writer has made the corrections.

This chapter also gave you information about how to report product-related problems. While product design issues and problems are not the documentation tester's primary concern, the tester should not ignore problems when they are discovered or assume that some-one else will take care of them. Documentation testers should do everything possible to help ensure the success of the product and the quality of the product documentation. Where appropriate, that includes reporting any product issues and following up until they are resolved or dismissed.

Clear, concise, properly prepared reports contribute to the success of the documentation test, and ultimately to the success of the product and its associated documentation. And you, as the documentation tester, are responsible for creating and distributing those reports.

References

Hetzel, Bill. 1988. *The Complete Guide to Software Testing*, 2nd ed. Wellesley, Massachusetts: QED Information Sciences, Inc.

Sides, Charles H. 1991. *How to Write & Present Technical Information*. Phoenix, Arizona: Oryx Press.

6

Becoming a documentation tester

To become a documentation tester, you need a good idea of what is expected of one. Knowing the responsibilities of a documentation tester is the first step in becoming one. Developing successful personality traits, acquiring the needed skills and abilities, and getting the proper education come next. This chapter discusses these relevant topics.

After reading this chapter, you will understand a documentation tester's

- Three primary responsibilities
- Additional responsibilities
- Successful personality traits
- Acquired skills and abilities
- Education alternatives

Three primary responsibilities

As with any job, there are certain tasks and duties for which the employee is responsible (Fig. 6-1). The primary responsibility of the documentation tester is that of planning and conducting tests on product documentation. Before a tester gets to the point of actually performing a test on documentation, however, there are two other tasks that must be completed. The successful completion of the documentation

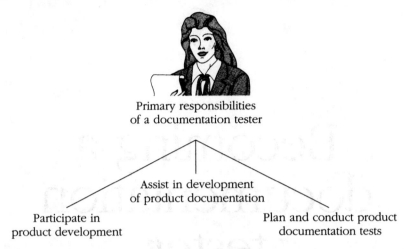

6-1 *Primary responsibilities of a documentation tester*

tester's primary task can be affected by the successful completion of these two other tasks:

- Participating in product development
- Assisting with documentation development

Participating in product development

The design and development of a new product can be both time-consuming and costly. The initial stages of design and development usually require many hours of meetings, many interoffice communications, many pieces of design documentation, and many design changes and modifications. The documentation tester's participation as a design and development team member benefits both the design of the product and the efforts of the product team's members.

Documentation testers often have a good understanding of the product's intended user. Unless the product is new, or the documentation tester is new to the product, a documentation tester brings a great deal of knowledge about the product's user to the design and development team. By the time a product team begins design review meetings, the documentation tester might already be heavily involved in several efforts related to the product and its defined user. Documentation testers often participate in end-user product testing, on-site visits to user's places of employment, and alpha or beta testing of the product. (An alpha or beta test involves the delivering of an otherwise finished product to a select group of typical users in an effort to seek their feedback and implement their suggestions in the final stages of product development.)

A documentation tester with this type of product and user knowledge can provide the design and development teams with the users' point of view. Significant improvements in product interface are often the result of a documentation tester's input. In addition, the tester contributes to product development and documentation development in several ways by:

- Improving testing preparation
- Assisting in documentation development
- Increasing developer awareness

Improve testing preparation

Having a documentation tester as a product design and development team member helps the tester become better prepared for testing product documentation. In many instances, the more the documentation tester knows about the product, the easier it is for the tester to conduct a documentation test. This is true for several reasons.

First, there might be some aspects of the product's documentation that cannot realistically be as thoroughly tested as they should be, or cannot be tested at all. If this is the case with documentation you are testing, the product knowledge that you gained by attending design and development meetings might have to be sufficient.

For example, product documentation for a pesticide might include instructions for inducing vomiting in case the product is accidentally ingested. As a documentation tester, it is your job to make certain these instructions are correct. If you were conducting a basic functionality test on this document, you would have to drink the pesticide, and then follow the instructions to induce vomiting. However, this is a bit extreme. Your participation in the development team should give you sufficient knowledge of this product to know the procedure is correct without having to actually test it on yourself.

Second, thorough knowledge of the product reduces the amount of time required to test the documentation. This is particularly true with products such as software.

Some software products have different screens that appear depending on a particular sequence of events. If you are already familiar with these screens, when the documentation states that a particular screen will open, a quick glance is enough to tell you whether or not the documentation and software are synchronized. If you are not already familiar with the different screens, you might have to spend added time determining whether the correct screen did open.

Third, thorough knowledge of the product can also improve the quality of the documentation test. Using the example of software products again, many screens look very similar. In fact, one effort that software designers and developers often consciously make is to keep a consistent look and feel for their customers. That often means that the layout and design of one screen is similar to that of another. If you know the different screens well enough to be able to readily differentiate between them, you are less likely to confuse one for the other when conducting a documentation test.

Knowing what to expect of the product also helps the documentation tester make better estimates of the time it will take to conduct the test. If the tester's estimates are far less than the actual required time, the tester runs the risk of creating a bottleneck in the schedule, interrupting and delaying the schedule, or working massive amounts of overtime to keep on schedule.

Assist in documentation development

Documentation testers assist the documentation development team in several ways (Fig. 6-2) by:
- Improving the writer's access to product information
- Providing a single reference for product information
- Representing the writer's interests in development meetings

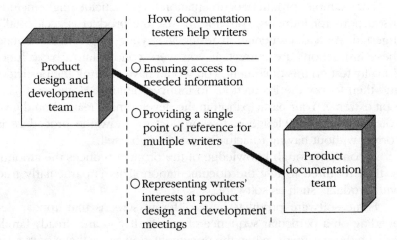

6-2 *Ways documentation testers help writers*

First, a documentation tester who has been a product design and development team member from the start can be a valued source of product information for the writer. Access to the documentation tester is often easier for the writer to obtain than is access to other design and development team members.

Second, the documentation tester can be a single point of reference for multiple writers. This not only ensures that all writers get the information they need about the product, but also reduces the interruptions developers would otherwise experience as the result of having to work with multiple writers.

Third, documentation testers can, because of their intimate knowledge of the documentation process as well as the development process, be a central representative for the documentation team on the development team. The documentation tester ensures that product developers are consistently aware of how changes they make late in the process can affect the product's documentation. The tester can point out areas in which the proposed modifications might not be worth the documentation and, subsequently, the product production delays.

Assisting in documentation development

Testing the product's end-user documentation is the documentation tester's primary responsibility. It seems logical that the documentation tester and the documentation can both benefit from the tester's participation in its development. As with the product's development, the tester should participate early in the development of the documentation. Early participation begins with the first documentation team meeting.

The documentation tester brings many skills and often much knowledge to the design and development of the product. The tester also brings these skills and related knowledge to the design and development of the product's documentation. For example, the documentation tester's knowledge of the user is important input in helping produce a product that meets the user's needs and wants. This is also true with the development of the product's documentation. The tester's input can help guide the documentation team in their efforts to develop user-friendly (easy for the user to understand and work with) documentation.

The most important way in which the tester can assist with the development of the product's documentation is by attending the documentation team meetings. During these meetings, the tester provides information as needed. Where possible, that information should help the documentation team to

- Determine the best approach to use when documenting the product

- Understand which product features are the most important to the user
- Make certain that information related to the user's most important features is not missed
- Correct any documentation problems that the user had with earlier versions of the document set, if earlier versions exist
- Develop a production time table that is realistic and does not delay the development and production of the product
- Ensure the finished documentation is of the highest possible quality given the team's time and cost restrictions

Both the documentation tester and the documentation team realize benefits from the tester's early and continued participation in the design and development of the documentation (Fig. 6-3).

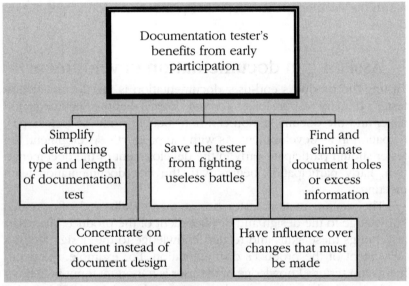

6-3 *Ways documentation testers benefit from early participation*

Benefits the documentation tester receives include

- Early familiarity with the product's documentation. This makes it easier and simpler for the tester to determine the most appropriate type of documentation test to conduct, and helps the tester determine how much time will be needed for the test.
- Complete concentration on document content. Issues such as document design are generally discussed and decided on during document and development design meetings. A tester

can save time during the test by providing input related to the document's design during development, instead of later in the process when modifications become costly and time-consuming.

- An understanding of why the documentation follows a specific design. This knowledge can save the tester many useless battles, allowing the tester to concentrate on the document itself.
- The ability to find and eliminate document holes as well as excess information. This knowledge gives the tester more time to look at areas where more information is needed in the document, or where too much information has been provided.
- Improve rapport with documentation team members. This rapport gives the writer a certain amount of influence over changes that need to be made. It makes it easier for the tester to suggest and help to implement late changes. In addition, if the writer feels comfortable with the tester, the writer might feel less threatened at having his work reviewed and scrutinized.

As mentioned previously, the planning and scheduling of the actual documentation test is easier when the tester is part of the documentation team. And conducting such tests is the documentation tester's primary responsibility.

Planning and conducting documentation tests

It is possible for a documentation tester to simply pick up the document and begin testing it, without any previous participation on development teams or previous knowledge of the product. In fact, each of us conduct such a test every time we read a piece of product documentation. In the process of reading documentation we often note errors and inconsistencies that the document's author did not see. It is the documentation tester's responsibility to catch these errors and inconsistencies before the document's reader does so.

To do a thorough job of testing product documentation requires that you first plan the test to be conducted. When planning for a documentation test, you must make three primary planning-related decisions:

- Determine which documents, or portions of documents, you will test
- Choose the type of test you will conduct
- Identify when you will start and finish your test

Determine what will be tested

You are most often called on to decide the last issue—when you will start and finish the test—before you can decide the first two. However, you cannot provide an accurate estimate of the length of time it will take you to conduct a documentation test, until you know what you will be testing and which type of test you must conduct.

Planning a documentation test requires that you define exactly what you will be responsible for testing. What you test in a given document depends on several factors:

- What you are attempting to accomplish with the test
- What specific issues or procedures you have been requested to test
- How much of the document is new
- How much of the document is changed
- How much of the document is unchanged from a previous version because the related product was not changed
- What unchanged parts of the original document were thoroughly tested the first time around

Choose the type of test

Not all documents require the same type or length of test. Your primary goal in testing the document determines the type of test you must conduct. For example, if your primary goal is to make certain that the conceptual information is accurate, you might choose to conduct a read-through test.

You might also choose the most appropriate type of test based on the desires, issues, or procedures that others have requested or specified. For example, writers sometimes have a greater difficulty obtaining sufficient information about one aspect of their document than another. If the writer indicates to you that this was the case with a given area of the document, you might want to conduct a more thorough test on just that portion of the document.

If the document you are testing contains only minor changes, the type of test you conduct can be different than if the changes were major. When the document to be tested is a previously released document, you must also consider whether or not the previous version of the document received a test. If the document received a thorough test originally, then testing only the changes might be sufficient. However, if there was only enough time last release to quickly scan or read through the document for gross mistakes, this might be the appropriate time to conduct a thorough test of the document.

Identify start and finish dates

Once you know exactly what you must test, what you should be looking for during that test, and what type of test you must conduct, then you can provide a reasonable estimate as to the start and finish dates for the test. Of course, providing this information is also dependent on other factors:

- How soon the document will be available for you to conduct your test
- What condition the document will be in when you receive it from the writer
- How much detail you must provide the writer when you locate problems in the document
- How many other documents you are required to test, particularly if the testing must be done simultaneously
- Whether you have other responsibilities associated with this document and product
- Whether a written report, a verbal report, or a combination of both is required

It does no good for you to schedule a documentation test if the document cannot be ready at the time you choose. Careful coordination with the writer becomes an important issue when determining start dates for your tests. It also becomes an important issue when determining the quality of the document you can expect to receive for documentation testing.

If the writer takes too much of the total development time writing the document, you will not have enough time to complete your test. On the other hand, if the writer cuts the writing time as much as possible, the quality of the document might suffer. This increases the burden of testing the document, which might also increase the length of time it takes to test it. There is a trade-off, and thus there is also a careful balance that both you and the writer should strive to achieve. This balance is reflected in not only the time required to complete the test, but also in the type of test you will conduct.

A document that is incomplete or contains many problems can grossly overload you with work related to testing the document. If you are aware that the document might have many errors or problems in it when you begin your test, then you can negotiate how much help you will provide the writer.

Sometimes documentation testers provide all information that is missing in the document, even writing it themselves if necessary. They might also conduct the related research and provide corrections to information that is incorrect. Other times, testers only note where information is missing or incorrect and leave it up to the writer to do the research, writing, and rewriting as needed.

Some documentation testers have other product responsibilities they must consider when negotiating the length and type of documentation testing, such as reporting and following up on product-related problems. Proper planning must, therefore, include time to handle these types of other product-related responsibilities.

In scheduling a documentation test, you also want to schedule time to report your results. Results can be reported directly to the writer in a meeting. You can also report them to the writer by preparing a formal written report. Often the most effective way to report the results is a combination of these two approaches. Whatever method you choose, some time and effort on your part are still required. You need to consider this when estimating how long it will take you to conduct your documentation test, and what the start and finish dates will be.

Planning for and conducting documentation tests is, as previously mentioned, the most important responsibility of a documentation tester, but certainly not the only one. In addition, documentation testers have a variety of other duties and responsibilities. Performing these other duties contributes indirectly to product quality by contributing to such things as the documentation testing process, the company, and the knowledge and skill level of other employees in the company.

Additional responsibilities

There are five additional responsibilities that a documentation tester should consider to be part of his regular job assignment (Fig. 6-4). The first four responsibilities come as part of the responsibility of working for the company, and in participating in the product's development. The fifth responsibility is often one that must be directly assigned to the documentation tester by someone else in the company. The five additional responsibilities of a documentation tester are

- Being a team/company player
- Working toward improving the process of testing documentation
- Advising, assisting, and mentoring other less-experienced documentation testers
- Advising and assisting other product design, development, and documentation team members
- Helping to develop and deliver product training materials and courses

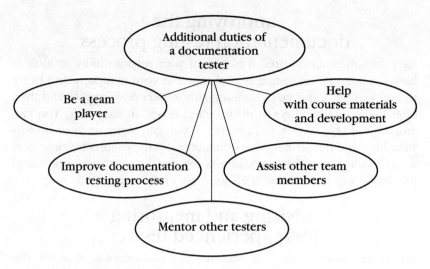

6-4 *The documentation tester's additional duties*

Being a team/company player

Being a team player in any company means doing your best for that company at all times. Your best might be described simply as getting to work on time each day and always doing the best you can in your job while you are at work.

More specifically, being a team player for the product design and development team, as well as for the documentation team, can have several requirements or definitions associated with it including

- Regular attendance at team meetings
- Contributing comments and suggestions to open discussions
- Helping out when and where needed, even if it is not part of your regular job description
- Maintaining a positive outlook
- Acting professionally and effectively
- Working toward the successful attainment of team goals
- Working well under stress and pressure
- Exhibiting a mature and helpful attitude and approach to problem solving
- Willingly working with others

Some of these requirements are expected of you even if you are not working in a team environment. For example, companies expect you to try to maintain a positive outlook and to act professionally and effectively in any assignment.

Improving the documentation testing process

As a documentation tester, it is part of your responsibility to look at how documentation testing is performed in your company and to try to find ways to improve it. Looking at how you do your job, and then objectively finding ways to do it better, is not an easy task. You can employ the help of others, however, if you can maintain your professionalism and not let negative comments or suggestions become personal insults or assaults. Ask others for their input. If you are asked for similar input related to their jobs, try to do your part as well.

Advising and mentoring less-experienced testers

The more experienced you become as a documentation tester, the more you can help to orient and train less-experienced testers. Sometimes, mentoring another tester can be done by reviewing a document they tested. You might see ways in which the document they tested can be improved. Point these out to the tester, but do so with respect for the tester as an individual. If you do so, not only will you improve your working relationship with this individual, but you will contribute to that tester's knowledge. This in turn benefits the product and the company as a whole.

Advising and assisting other product team members

Let other team members know that you are available to help if you can. Provide advice with problems or issues if your advice is sought. There might be occasions where you can provide advice when your input has not been openly asked for. Part of building and retaining team rapport is the skill of helping at the right time—when people need and want it.

Helping to develop and deliver training materials and courses

Developing and producing training materials is another area in which a documentation tester can be particularly useful. Sometimes the only effort that a tester must put forth in this area is that of working with a course developer. Reviewing course materials for technical inaccuracies, providing information as needed, or simply helping the course developer get in touch with the right person to answer questions are all ways in which you can help develop training materials.

Another way to help with training materials is writing them, if you are specifically requested to do so. However, it is not common for documentation testers to write course materials.

In addition, documentation testers sometimes teach training classes. They might also participate as a student in prototypes or early product training classes, providing useful feedback to the course writers and instructors.

Successful personality traits

As with most jobs, some people are more skilled at one type of job than they are at another. The set of skills that any one person brings to a given task are most often the result of acquired knowledge and/or practice. However, some people seem to have a knack for doing some tasks better than others. That knack can often be associated with various personality traits.

Do not assume, however, that just because you do not possess all of these traits that you cannot be successful as a documentation tester. It simply is not true. Your contributions as an individual to a team of testers is equally as valuable as that of each of the other testers. It is the team together, including their diversities, that make for strong and effective testing. If you do not possess all of the personality traits discussed here, concentrate on learning the skills that will help you become an active and contributing member of the documentation testing team.

Several of the personality traits (Fig. 6-5) that contribute to your effectiveness as a documentation tester are

- Group participation skills
- Communication skills
- Product aptitude
- Ability to handle stress
- Ability to adapt to change
- Organizational skills

Group participation skills

As a documentation tester, you can easily spend up to half of your time attending meetings or otherwise participating in group activities. You will be regularly attending product design and development meetings, as well as documentation meetings. You will probably also regularly attend documentation testing team meetings. You will be expected to participate in, not just attend, each of these meetings.

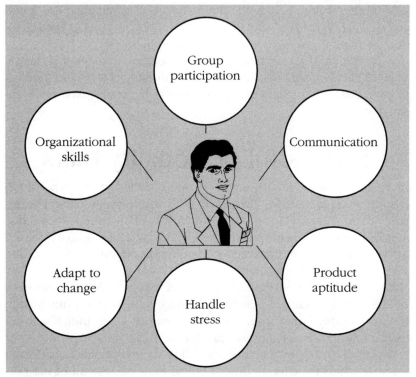

6-5 *Personality traits of an effective documentation tester*

Meeting participation involves asking questions to clarify information. It also includes sharing information with other team members. Sometimes you might be called on to prepare and deliver brief presentations. You will also find yourself distributing copies of and soliciting feedback on testing schedules and other documentation testing-related documents.

Personality traits that you will find particularly effective when group participation is necessary include

- Being a good listener
- Always being prepared
- Having a sense of humor
- Being empathetic
- Being outgoing

Being a good listener

This personality trait or skill is important because you will have to do a great deal of listening in product design and development meetings, as well as in documentation meetings. These meetings are where you will obtain much of the product knowledge you will need in order to successfully understand and test the product documentation.

Good listening skills require that you hear with your mind and emotions, as well as with your ears. Listen not only to what is said in meetings, but to what the speaker means. Clues to the speaker's unstated meaning can be found in how the speaker says what he says, as well as how he looks and acts when he says it.

Listening skills are also useful for hearing what is being said even when no one is speaking. For various reasons, most often the fear of job loss, many meeting participants are reluctant to state what they really believe or feel. Being able to hear what is not being said can provide you with important clues about the product. With this skill, you might be able to see aspects of the product's related documentation on which you should direct your concentration when it comes time to conduct the documentation test. At the very least, you can note individuals with whom a one-on-one meeting might later provide you with more product information.

Always being prepared

When you attend a design or documentation meeting, come fully prepared to discuss issues that concern you, as well as to learn about the product. Coming fully prepared to meetings has several meanings.

First, be physically prepared by bringing paper, pens, your laptop computer, whatever you need to take notes. Second, show you are fully prepared by arriving on time for the meeting. Being on time shows respect for others and an interest in the product. Third, be prepared by previewing any documents distributed before the meeting. If they raise questions in your mind, note them so that you can ask those questions in the meeting. Do not simply rely on your memory. Answering questions is one of the main functions of these meetings. If you forget to ask your questions during the meeting, you will have to ask them later. Finally, be prepared if you have an active part in a meeting, such as distributing a testing plan or making a presentation. Be certain that any equipment you need for making a presentation is available and functioning. Also, make sufficient copies of any documents to be distributed, or be prepared to tell those attending the meeting where copies can be obtained.

Having a sense of humor

No matter how prepared you are, how professional you act, how much concentrated effort you put forth, things do not always proceed as planned. Be flexible when things go wrong. But most of all, keep your sense of humor. Take your job seriously. Always make a serious effort to do the best you can. But when nothing you do helps, find some humor in it. Humor helps to relieve the tension for you and others, and clear your head. Then you can get back on the right track.

Being empathetic

In addition to having a sense of humor, it is important to know when you should be empathetic. Team members are people. People have problems. Sometimes the problems that arise might not directly affect you. Sometimes, however, problems can interfere with the effective functioning of the team. In these instances, a little bit of empathy can be helpful in normalizing the situation, letting the other person overcome the problem and moving the whole team forward once again.

Being outgoing

Establishing a good working relationship with team members means you must also learn to be friendly toward them. Try to get to know a little about each of the other team members. As time passes, you will find that cooperation and assistance are a lot easier to come by if you have first established a friendly relationship with other team members.

Communication skills

Two types of communication skills are key to being a good documentation tester—verbal and written.

Verbal

Verbal skills involve the ability to talk with other people in such a way that you effectively communicate the information you intend to communicate. Verbal communication skills also involve being able to ask the right questions, using the right words, so that you get the answer you were looking for, as well as being able to listen effectively.

If you are uncomfortable with your present verbal skills, you can work toward improving some aspects of your verbal and presentation skills. Some areas in which you can improve your skills include

- Preparation
- Voice control
- Eye contact
- Distractions
- Pace
- Enthusiasm

Always come prepared to discuss the topic at hand or to ask questions. If you need to make a presentation, make certain you know your material. Practice it until you feel comfortable with it.

Also, control your voice so that its pitch is neither too high nor too low. Occasionally modulate the tone and enthusiasm portrayed in your voice so as to provide variety.

Keep eye contact with your audience whenever possible. It lets you see anyone who wants to ask questions. It also lets you determine, by the look on their faces, whether they understand the material you are presenting. Keeping eye contact with your audience also helps you portray a sense of self-confidence.

If you have habits, such as always repeating the same words (you know, is that clear, uh, etc.) or jingling change in your pocket, make an effort to avoid these habits during your presentation.

Deliver your presentation at a pace that is comfortable for you, as well as for your audience. If the audience is not catching on to what you are explaining, perhaps you are speaking too fast. Slow down a bit. If the audience seems to be dozing off, pick up the pace a little.

Make an effort to reflect in your presentation, your voice, and your mannerisms the enthusiasm you feel for the product or topic you are discussing.

If it is too difficult for you to improve in these areas by yourself, consider taking a speech class at your local adult education campus or community college. You can also consider joining an organization such as Toastmasters that concentrates on helping you and others improve their presentation skills.

Written

Written skills include not only the ability to write interoffice communication and documentation testing plans, but also the ability to write exactly what you mean when you make notes on the document you are testing. The document's writer needs to know exactly what you found wrong, how it is wrong, and if possible, what suggestions you would make for correcting the problem. That often means that you must write a suggested sentence, paragraph, or even several pages of documentation.

Writing requires that you be able to communicate information in a logical, sequential, clear, and concise manner. You should be able to use correct grammar and sentence structure as well. However, your writing does not need to be so formal that it is unreadable. You should write in a straightforward and friendly manner. The point is to get the idea across to the reader with as few errors and problems as possible.

If your writing skills need improving, you can purchase a book to help you with your grammar and writing. You can also attend a course at your local adult education campus or community college.

Another effective method of improving your own writing is that of reviewing writing done by others. Read through it and try to figure out what makes it a good piece of documentation. Use magazines, technical documents, books, and newspapers to introduce yourself to a variety of writing styles. Diagram some of the sentences that you find in these resources. (Any good English composition book can explain sentence parts and diagramming to you.) Use a good style guide, such as *The Chicago Manual of Style*, to help you understand why the layout and style of an article or book is effective.

Product aptitude

It never hurts to have an innate understanding of the product with which you are working. It is not mandatory, however. You can develop an aptitude for the product for which you are testing documentation by working with the product. Some people will catch on quicker than others. If you are diligent, you will begin to understand (develop an aptitude for) the product whose documentation you are testing.

Although developing an aptitude for some products requires more effort than others, your documentation testing skill will improve as you develop your understanding of the product. Therefore, putting time into this aspect of being a documentation tester is important.

Ability to handle stress

Nothing ever goes smoothly. That statement is more true in some environments than it is in others. For example, I have often felt that in computer software documentation testing—the field with which I am most familiar—Murphy could write an entire set of his laws.

Knowing that nothing ever goes exactly the way you plan for it to go should make it less stressful for you when glitches occur. Unfortunately, problems still cause stress. There will be more stress associated with some documentation testing products than with others. The key is to be able to handle the stress.

Realize that no matter how hard you try, problems are inevitable. Accept problems for what they are—a chance to improve or at least change something. Changing the way you look at problems can be a big step toward controlling stress.

Do not let problems get blown out of proportion. Once you discover a problem, make every effort to solve it before it becomes too big to handle. Get everyone whom it might affect involved in its solution. Working together, you should be able to solve problems before they become too stressful to deal with.

However you approach it, handling stress is an important part of becoming a good documentation tester. You must learn to handle it, because you often cannot avoid it.

Ability to adapt to change

How wonderful it would be if, once you learned how to do something, it never changed. It would be a little like riding a bicycle. After years of neglect, you might be out of practice, but you still remember the procedure because the procedure has not changed. Well, that might be true with bicycle riding, but it is rarely true with much else, particularly with documentation.

Documentation changes frequently. Changes to documentation require additional documentation testing. In a way, the inevitability of change provides you with a small amount of job security. But usually, the changes that occur with documentation have less effect on you than the everyday changes that occur around you.

Some changes you can expect to encounter include things such as new personnel joining the development and documentation teams, deadlines being moved up, and late deliveries of documents for testing. No matter what the change is, you need to be able to adapt to it. Taking changes in stride and working with them the best you can shows your professionalism and dedication to the product.

Expect change. It will come. Prepare for it mentally, accept it when it does come, and your life will be much easier.

Organizational skills

There are four organizational skills that will prove to be particularly useful in your daily activities as a documentation tester:

- Meet deadlines
- Be a self-starter
- Enjoy learning
- Enjoy problem solving

Ability to meet deadlines

If you can meet deadlines, even when everything around you is fighting against it, you will have successfully developed your organizational skills. Meeting deadlines requires dedication, hard work, long hours (sometimes), and a desire to meet those deadlines no matter what.

Meeting deadlines requires one other skill—the ability to schedule and organize your workload. Scheduling and organizing your workload so that you can meet your deadlines means understanding what is involved in the task you must complete. Understanding what is involved in testing even a single page of product documentation gives you a good head start on effective scheduling and organizing.

The next important step is to use your knowledge and not let others dissuade you from your schedule. You will have to make other team members understand that if it takes you two hours to test one page of documentation, it takes two hours to test one page of documentation. You cannot change that time requirement unless they are willing to give something up in return.

To reduce a time requirement, you must either reduce the quantity or quality of work or increase the related resources. In other words, others must be willing to have fewer pages tested, a less-stringent test performed, or more testers added to the project.

When you schedule documentation tests, schedule them and organize your time so that you can meet the schedules you set. Then, only changes requested by others should affect your schedule. If others request changes, your careful organization and scheduling should help them to see what they must give up (quality or quantity) or what they must contribute (additional resources) in order for you to accommodate their request.

If you consistently meet your deadlines, people will come to trust your knowledge and your schedules. They will eventually come to know that if you say it will take three days to complete a test, it will take three days, and that only their cooperation, sacrifices, or contributions of additional resources make it possible to reduce that time.

Be a self-starter

Procrastination is not uncommon. Getting started on a documentation test is often the hardest part of conducting one. You must be an individual who is capable of getting themselves going, without someone or something forcing you into it.

This does not mean that if you have occasional procrastination episodes that you should not be a documentation tester. It is what you do when you encounter these days that matters. If you do in fact procrastinate until all hope is lost, consider another type of job. If, however, you get yourself out of that rut and get the job going, then you have enough self-starting ability to be a good documentation tester.

Enjoy learning

Above all else, you must truly enjoy learning. As a documentation tester, you are likely to learn something new every single day. If you enjoy this type of an environment, your job as a documentation tester will be pleasurable. If the opposite is true, you might find yourself dreading each day at the office.

Enjoy problem solving

Since nothing ever goes smoothly, you must also enjoy problem solving. If you can see each new problem as a challenge, rather than as a roadblock, you will be well on your way to becoming a good documentation tester.

Acquired skills and abilities

Most of the personality traits just discussed are traits that develop over time. You might be able to improve your abilities in some of these areas; others you will just have to live with. There are other skills and abilities a good documentation tester needs. These are acquired through effort, experience, practice, and education. Some of the skills and abilities that will prove useful to you as a documentation tester include

- Product experience
- Writing skills
- Editing skills
- Business sense
- Industry knowledge
- Computer aptitude

Product experience

Your ability to detect problems and errors in product documentation will improve as you become more knowledgeable of the product whose documentation you are testing. Having experience with the product before you become a documentation tester, however, is the type of product experience that will benefit you in other ways.

The main benefit is the intimate knowledge of the product's user that it affords you. You become a better tester because you already have an understanding of the product's strong and weak areas. You can then use this experience and knowledge in several ways.

As a product user as well as a team member, you can point out specific areas where improvements in the product would benefit the product and those who use it. You can also make suggestions for product changes and improvements based on your own experiences with the product.

When you participate in product documentation meetings, you can point out areas where the documentation can be improved. Sometimes you might have suggestions for reorganizing the documentation. Sometimes you might have suggestions for changing the type of information in the manual. Sometimes, you might suggest that no changes be made at all. Knowing what has been done right, as well as what can be improved on, is beneficial to the entire documentation set.

Writing skills

As a documentation tester, you might find it necessary to write portions of the manual you are testing. One of your responsibilities is to find holes (areas where needed information is missing) in the documentation. Once you find these areas, you write in the information. It is then up to the writer to use what you wrote with some modifications, use it exactly as you wrote it, or completely rewrite it. Regardless of the writer's decision, you still need to have enough writing skill to let the writer know what information is missing.

Sometimes you will have to rewrite portions of the manual you are testing. If the writer has not accurately explained a process, concept, or other technical information, you will need to rewrite the inaccurate information. Good writing skills are needed in this instance as well.

Even if you never have to change a word the author has written, you still have to write reports. You might not need to write more than one page to explain what your documentation test discovered about a piece of documentation. But even that one page needs to be professionally written.

You will also find that it is a lot more difficult to gain acceptance from writers, if you yourself do not also have adequate writing skills. Look at it from the writer's point of view. Would you feel comfortable having someone test the writing you slaved over, if you knew this individual did not have basic writing skills?

Editing skills

You also need to have basic editing skills. While you will sometimes be required to write portions of a document you are testing, more often than not you will need to rewrite small portions. Basic editing skills will prove particularly valuable in these situations. When you rewrite a portion of a document to correct technical inaccuracies, your rewrite must be legible and must accurately state the information it is intended to convey.

Business sense

Few jobs are done just for the sake of doing the job. Most often, jobs are done to accomplish a specific business objective, that of earning a net profit. Remembering the company's business objective, and working toward meeting that objective can be referred to as having a corporate or business sense.

When testing a document, keep in mind the company's objectives related to this document, including the effect your testing has on the quality of the product, and therefore on the company's net profit. It will help you focus on what is really important in the document, where and when changes and corrections are essential, and when your emphasis should be changed to encompass the company's goals.

Industry knowledge

Understanding the industry of which your company is a part, and the competition for your company's product, helps provide you with a solid foundation for testing documentation. Few products have no market competition. Understanding the pros and cons of competing products and their associated documentation helps you see areas in your own product and product documentation that can be improved.

Understanding the industry also means understanding other factors that have an effect on your company's product. Some of those factors are

- Product packaging. How your product is packaged affects how much and what type of documentation you can include with your product. Some types of packaging lend themselves to more traditional documentation—user manuals and booklets. Other types of packaging give you the opportunity to experiment with the documentation. Others can severely limit how you provide documentation.
- Sales seasons. Whether the product you produce sells all year long, or only sells at certain times of the year, can also affect your documentation. For example, if your company produces a product that sells all year long, the documentation you include in this product must be of sufficient quality that you will not need to update it regularly. If you create a product that only sells at Easter, the longevity of the documentation might not be as important.
- Indirect competition. Other nonsimilar products can compete with your product. For example, companies that produce typewriters have not only other typewriters as their competitors, but must see word processors, computers, and to some extent, printers as their potential competition. You can benefit by becoming familiar with these products and their approaches to product documentation.

To develop a knowledge of your company's industry, read trade publications for your industry. If possible, examine the competing products. Read the sales literature other companies put out, as well as your own company's sales literature. Use other products. Read their documentation. Try to find what the competition does better than you, then apply that knowledge to improve your product and documentation.

Computer aptitude

Computers are used in so many businesses today that you really cannot get by without knowing how to use one. At the least you should know how to run one type of computer. You should probably also learn at least one type of application program, such as a word processing program. Even if your company does not work with computers in the production of its product, you will probably use one for report writing, if nothing else.

Education and training alternatives

One of the more interesting aspects of being a documentation tester is the variety of documentation that you can test. As a documentation tester, you can test documentation for software application programs just as readily as you can test documentation for deep sea diving equipment. Any field that has documentation associated with it can use documentation testers.

I must add a caveat here, however. Just because documentation exists, and even though the documentation might need testing, does not mean that documentation testers are routinely hired. The truth is that some fields use documentation testers and some do not. One field that makes use of documentation testers is computer science. Both software and hardware companies use documentation testers to help improve their documentation, although they might not apply that title to the position.

Once you determine whether documentation testing or its equivalent is used in the field you prefer, you have a variety of options available to you for gaining the skills you need to become a documentation tester.

Education alternatives

Education generally refers to formal education obtained attending a college, university, or technical school. Colleges, universities, and technical schools provide alternatives in education—both in the level of education as well as the field.

There are different levels of education from which you can choose. Levels include certificates, associate degrees, bachelor degrees, master's degrees, and doctorates. You should choose the level that best meets your needs and resources.

For example, if you are interested in becoming a documentation tester in the field of physics, you are likely to need a minimum of a master's degree in physics. On the other hand, if you are interested in doing documentation testing for software development, a bachelor's degree in computer science might be the highest degree you need.

Before you decide what level of education is sufficient to obtain a documentation testing job in your field of interest, consider which field interests you the most. Research the education alternatives and employment opportunities for that field. One reference book that you might find useful is *The 1994 Information Please Business Almanac and Desk Reference* (Godin 1993). Your local librarian should be able to assist you in finding other books and information as well.

Education fields

Some of the fields where the services of a documentation tester are more likely to be needed include
- Biological (life) science
- Computer science and information systems
- Electronic and computer technology
- Engineering science
- Environmental technology
- Finance and banking
- Machine tool technology
- Physical science

Biological (life) science
This field encompasses many fields related to the living world. Studies in the biological sciences include courses in anatomy, biology, botany, ecology, genetics, microbiology, nutrition, physiology, and zoology. A degree in this field will help prepare you for jobs in companies that deal with hospitals, research laboratories, or other related biology-oriented facilities.

Computer science and information systems
This field includes computer science, information systems, networking, electronic and computer technology, and other related fields.

Studies in computers and information systems include courses in computer basics, DOS, applications, programming, business, databases, data communication, systems analysis and design, graphics, and business math. A degree in this field will help prepare you for jobs in companies that deal with computer equipment, computer software design and development, robotics, information management, and other fields related to computer science and information systems.

Electronic and computer technology
Though considered a field in itself, electronics and computer technology is closely associated with the computer science and information systems field previously mentioned. However, this field concentrates more on electronics and circuitry than computers. Studies in electronics and computer technology include courses in dc/ac circuits, digital and analog devices, microprocessors and computer systems, communication systems, and other related areas. A degree in this field will help prepare you for work in companies that manufacture electronic equipment, robotic devices, communication equipment, defense products, and other electronic products.

Engineering science
There are a wide variety of engineering science areas including chemical, civil, mechanical, computer, electrical, manufacturing, biological, and aerospace. Studies in the engineering sciences include courses in mathematics—including algebra, calculus, and geometry—and engineering—including civil, chemical, and electrical. A degree in this field will help you prepare for a position with a company specializing in any of the above-listed fields, as well as areas of specialization such as architecture, nuclear engineering, petroleum mining and production, and others.

Environmental technology
This field includes sanitation, public health, water purification, and other environment-related technologies. Studies in the field of environmental technology include courses in wastewater treatment, environmental law, aquatic microbiology, corrosion technology, and other related subjects. A degree in this field will help you prepare for a position in companies responsible for the treatment of wastewater, the design and manufacture of wastewater and other recycling equipment, and other related companies.

Finance and banking

This field includes anything related to financing and banking. Studies in the field of finance and banking include courses in business, marketing, supervisory management, banking, finance, organizational behavior, economics, business law, business communication, and other related courses. A degree in this field will help you prepare for work with a company such as a lending institution, the federal government, or other related companies.

Machine tool technology

This field encompasses many types of businesses. Almost any company involved in manufacturing uses machine tool technology. Studies in the field of machine tool technology include courses in mathematics, machine tool measurement, punch and die, manufacturing, arc welding, plastic mold making, metallurgy, and other related courses. A degree in this field will help you prepare for a position with a manufacturing company, including companies that manufacture machining tools.

Physical science

This field encompasses a variety of courses related to nature, the physical world, and the universe. Studies in the field of physical science include courses in astronomy, geology, chemistry, physics, and other related courses. A degree in this field will help you prepare for a position with companies that use the physical sciences, particularly those in research and development, both in the private and public sectors.

Education levels

Once you have decided which field is of the most interest to you, your next decision is the level of education you want to obtain. To determine that, you need to find out what level of education employers generally require. Determine that information by contacting the personnel departments of different companies in your field of choice and ask them about the minimum education requirements for related positions. You can also check with your local reference librarian for books and periodicals that might provide you with the information you need. The more common levels of education available to you are

- Certificates
- Associate degrees

- Bachelor degrees
- Master's degrees
- Doctorates

Each of these levels requires different quantities of knowledge, and sometimes different approaches to obtaining that knowledge. For example, you can obtain certificates from several sources such as private training companies, private colleges, community colleges, or state universities. Attending a community college, a private college, or a state university is often the most common way of obtaining any of the degrees previously listed.

Certificates

Certificates are offered in a variety of fields, although the most common fields are business related or technical. For example, you can earn certificates in business management and CAD (computer-aided design). The life sciences and other similar fields usually offer an associate degree as a minimum, rather than a certificate.

Requirements for obtaining a certificate vary almost as much as the types of certificates you can earn. It is possible to earn certificates for as little as four hours of course attendance. Colleges and universities frequently offer certificates for one-year courses. One-year certificates are offered in fields such as

- Accounting
- Business management
- Computer science and information systems
- Environmental technology
- Finance and banking
- Machine tool technology
- Office technology

Associate degrees

Associate degrees are offered in most fields. There are two types of associate degrees—science and art. Associate degrees require the equivalent of two full-time years of college work. In general, to obtain an associate degree you must have the following:

- A total of 64 semester credits (or equivalent).
- Overall grade point average of 2.0 (C).
- A minimum number of credit hours earned in attendance at the degree-granting institution.
- Completion of core course requirements as determined by the institute granting the degree. This usually requires a variety of

English, math, history, economics, humanities, social science,
biological science, and physical science courses.
* A minimum number of credits in an academic or
  individualized program (area of specialty).

Bachelor's degrees

Bachelor's degrees generally require that you first meet the requirements for an associate degree, even if you choose not to actually get
the degree itself. In addition to meeting the requirements for an as-
sociates degree, to get a bachelor's degree you must also meet other
minimum requirements.

The requirements that you must meet to obtain a bachelor's degree vary from college to college, as well as from one type of bachelor's degree (bachelor of arts) to another (bachelor of science).
However, in general you must complete a minimum number of credit
hours, somewhere around 180. This usually means an additional two
years of college work, on top of the two years it took to obtain the associate degree. Of those credit hours, some require that you take a variety of general education courses such as arts and humanities. Other
requirements relate directly to the subject area in which you are majoring.

The credit hours you seek must come from approved course
work. Courses that you take at one university might not meet the requirements of another. Therefore, you should check on the acceptance of course work between universities if you will be starting your
degree program at one university, but finishing it at another.

In addition, you sometimes must have permission from the chair-
person of the department in order to enroll in a degree program. Again,
these types of requirements vary depending on the department/divi-
sion, as well as the college or university. Before choosing any course
of action, check with the college or university of your choice.

Master's degrees

As with bachelor's degrees, master's degrees are also available for a
variety of specialties. In addition, there are two types of master's degrees—master of art and master of science. The type of master's degree you seek generally depends on your program of study.

Master's degrees have several prerequisite requirements. You
might have to prove proficiency in language competence. You can
usually prove your proficiency by taking and passing, with a B or better, a third-quarter language course. Another alternative is to pass the

Graduate School Foreign Language Test (GSFLT) with a sufficient score, usually 450 or higher.

Doctorates

Doctorate degrees are usually awarded after you have received a master's degree and fulfilled additional requirements. However, some doctorates are awarded without first receiving a master's degree. For example, the M.D. (Medicine Doctorate) does not require that you first receive a master's degree, although you generally earn sufficient college credits to do so.

To be admitted to a doctoral program, you have to meet requirements such as approval for admission, course work completion, successful completion of a comprehensive examination, and other requirements, depending on the doctorate you are seeking.

The doctorate degree generally carries with it requirements for such things as a minimum residency, the passing of an oral examination, a dissertation, and other requirements. In addition, there is also usually a time limit within which all requirements must be met.

Obtaining a degree or certificate requires a commitment of time and resources. You do not always need a degree to be successful as a documentation tester. You will find that in some fields, while a degree might help you get your resumé looked at, your experience is more important than the degree.

If you do not have a great deal of experience, or have not yet learned how to be a documentation tester, you have some options for gaining training and experience.

Training and experience options

There are several ways to gain the knowledge and skills you need to be successful at any job, including that of documentation testing. Two of the most common methods of gaining experience are

- On-the-job training
- Related experience

On-the-job training

You can learn the job that you want to do by taking on an apprenticeship type of position that provides you with on-the-job training. Many businesses work with colleges and universities to provide students with practical experience. These programs are often referred to as *internships*.

Internships can run for any specified length of time. Generally they run for a semester, an academic year, or a summer. Different businesses and universities establish internship programs to meet their needs, as well as those of the student.

Students enrolled in internship programs work part- or full-time hours while they continue taking courses. Internship candidates are usually in their senior year of college, although each college sets different requirements. Many interns perform the same job duties as the company's full-time employees. However, interns are generally supervised not only by a business supervisor but also by a college instructor or professor. These internships are intended to give the students an idea of the practical use to which they will be able to put their skills once they have completed their college education.

Other types of on-the-job training are available besides the internship programs that businesses have established in conjunction with colleges and universities. For example, part-time and volunteer work can also provide a substantial amount of on-the-job training.

Related experience

Although you might be seeking a position as a documentation tester, you do not necessarily have to have experience as a tester to be considered for such a position. Sometimes the experience you gain in a related field or position is equally valuable. For example, in software development, an individual who has programming or software testing experience might prove to be an excellent documentation tester.

You can see this type of relationship in other fields as well. For example, many of today's sports announcers are retired athletes. The point is, if you have experience, knowledge, or skill in a field related to the field in which you want to do documentation testing, then you can add value to both the company and its product. You should, therefore, consider those experiences and that knowledge when considering documentation testing.

Summary

The documentation tester has several responsibilities which, if properly fulfilled, can contribute substantially to the quality of the product documentation. The most important of these responsibilities is that of planning for and conducting the documentation test, the accomplishment of which is supported by the tester's efforts in two other areas of responsibility—attending team meetings and participating in the design and development of both the product and the documentation.

The documentation tester has an assortment of other responsibilities as well, some of which are common to different companies and job titles. For example, it is the responsibility of all employees to assist other less-experienced employees to become better at their jobs (mentor other testers) whenever possible. Some responsibilities are more specific to the job assignment itself, such as working to improve the documentation testing process, and helping to create and deliver product-related educational courses and materials.

To be a successful documentation tester, each of us must realize and accept that we are unique individuals who bring a variety and combination of talents, education, experience, and personality traits to the tasks we perform. While some of those skills, experiences, and traits are more useful to a documentation tester than others, all of a person's talents and skills are important. Some of those skills and talents are a matter of refined personality traits, such as a willingness to work in a team environment. Part of those skills and talents, however, are acquired through education and training.

To get the education and training you need to be a successful documentation tester, first look at your own past experiences and your current education. Determine if any of your knowledge and skills can be applied to documentation testing. Once you determine the area where your knowledge might be most useful, then begin looking for documentation testing positions in the same or related fields. If you decide that you need additional education or experience, there are alternatives. You simply need to research those alternatives and choose the one that best fits your needs and goals, then pursue any appropriate education or training.

However you approach your goal of becoming a documentation tester, remember there are many roads leading to success in such a job. Explore your options, then choose the one best suited to your talents, experiences, and desires.

References

The Chicago Manual of Style, 13th ed. 1982. Chicago: University of Chicago Press.

Godin, Seth, ed. 1994. *The 1994 Information Please Business Almanac and Desk Reference*. Boston: Houghton Mifflin, pp. 320–374.

7

Starting a documentation testing group in your company

Chapter 6 explained what personality traits, training, and experience contribute to the making of a good documentation tester. Your aspirations, however, might go beyond that of being a documentation tester. You might be interested instead in starting, developing, and managing a documentation testing group in your company. This chapter concentrates on providing information to help you achieve that goal, if that is what you choose to do.

To start, develop, and manage a documentation testing group in your company means you have to begin by proving the merits of documentation testing to your company's management personnel. If you are not already a member of the company's management team, it might be necessary for you to be the company's documentation tester for awhile, in addition to performing your current job duties, before you will have the opportunity to convince the company of the merits of a documentation testing team. On the other hand, if you already perform management duties and believe the addition of a documentation testing team is justified, you might be in a position to develop, document, and sell the idea of documentation testing to your company.

Assuming you are prepared to begin the process of developing, documenting, and selling the idea to your company, this chapter is the place to learn one approach you can take to help ensure the success of your idea.

After reading this chapter, you will understand

- How to overcome potential problems commonly encountered when proposing any new business idea
- How to develop a management proposal
- How to effectively present your proposal

Overcome problems

You are likely to encounter resistance when you attempt to convince your company that documentation testing is a reasonable, cost-effective method for improving the quality of their product documentation. Management is likely to be skeptical. For that matter, so is just about everyone else, until you prove that documentation testing is, at the very least, worth looking into. But you cannot even begin to prove the merits of documentation testing for your company until you accomplish three tasks (Fig. 7-1), the first of which is overcoming problems. The other two are creating a management proposal and delivering a presentation of that proposal.

The first potential problem—getting approval to research the idea of starting a documentation testing group in your company—is your first challenge.

7-1
Tasks to accomplish to overcome problems

Get approval to research the idea

If your company is one that actively seeks solutions to problems and solicits ideas and suggestions from its employees, you are already one step ahead of the game. A company that seeks solutions or solicits suggestions is usually a company open to change and improvement, regardless of who it comes from. Therefore, the first step you should take is to determine if your company is such a company (Fig. 7-2).

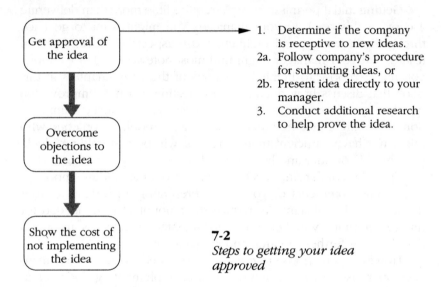

1. Determine if the company is receptive to new ideas.
2a. Follow company's procedure for submitting ideas, or
2b. Present idea directly to your manager.
3. Conduct additional research to help prove the idea.

7-2
Steps to getting your idea approved

Companies that seek problem solutions and ideas often have a system in place to encourage employees to contribute. Some companies establish suggestion boxes, or have committees set up to accept ideas and suggestions. Open forums might be held on a routine basis to which all interested employees are invited. Brainstorming (the free exchange of ideas without fear of reprisal, reprimand, or insult) is another method of offering problem solutions and ideas. Whatever method your company has established, consider taking advantage of the established procedure to make your recommendation for documentation testing as a method of quality control and cost containment.

If your company does not have a system in place for dealing with ideas, suggestions, and problem solutions, you need to seek a more traditional route. It might be necessary for you to present the idea directly to your immediate supervisor, and hope that it gets passed on up the chain to the highest required level.

Assuming that one or both of these options are available to you, you must still gather information and prepare some type of written report or verbal presentation in order to obtain permission to pursue the idea of implementing documentation testing in your company. Later in this chapter, you will find information on how to prepare a management proposal. Before you get to that stage, however, it is necessary for you to research the benefits and drawbacks of documentation testing as it relates to your company.

Getting initial permission might require little more than delivering a copy of this book to your manager. You might want to go back through it, however, and highlight those aspects of documentation testing that your manager might find most noteworthy. For example, you might choose to highlight sections of the early chapters in this book that discuss how documentation testing can help improve the quality of your company's product, as well as its product documentation. If you are concerned that presenting this book to someone who might not have sufficient time to read it will be a stumbling block, you should consider another approach.

Consider condensing this book into an outline, adding important quotes and comments of your own. Presenting an outline to your manager might substantially reduce the amount of time needed for him to determine whether or not documentation testing is an option that deserves further research and explanation.

However you choose to pursue it, you need to obtain your manager's permission to research the idea of implementing documentation testing in your company before you spend large amounts of time doing so, with one exception. If you choose to do all of the research, preparation, and so on during your off-work hours, you can do so without requesting permission. The fact that you spent your own time and funds instead of the company's to research this idea might even make it easier for you to get permission from your manager to continue developing the idea and possibilities associated with documentation testing. After all, if you feel strongly about an idea, and are willing to put in your own time and energies to pursue the idea, management is more likely to look at your suggestions more seriously.

Regardless of the approach you choose to pursue, your goal is to receive approval to research and develop the idea of documentation testing in your company. Once you have overcome the problem of getting approval to research the idea of documentation testing for your company, your work has just begun. There are many other problems to overcome.

Four additional problems (Fig. 7-3) you can expect to have to overcome if you are to be successful at implementing documentation testing in your company are
- Resistance to change
- Cost of not implementing documentation testing
- Writers who do not want anyone to review their work
- Engineers who are uncooperative

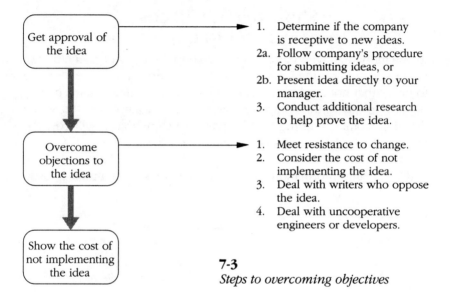

Get approval of the idea

1. Determine if the company is receptive to new ideas.
2a. Follow company's procedure for submitting ideas, or
2b. Present idea directly to your manager.
3. Conduct additional research to help prove the idea.

Overcome objections to the idea

1. Meet resistance to change.
2. Consider the cost of not implementing the idea.
3. Deal with writers who oppose the idea.
4. Deal with uncooperative engineers or developers.

Show the cost of not implementing the idea

7-3
Steps to overcoming objectives

Meet resistance to change

Change is a continual process that many people dislike. Change is often difficult and painful. Change sometimes means a loss of respected and valued things such as free time, the ability to make one's own choices, and so on. It is the rare individual who accepts change willingly and without apprehension or doubt. Change is stressful, and few people would deliberately add stress to their lives if given the choice. However, if you are going to implement documentation testing in your company, there is no way to avoid making changes. Consequently, one of the more likely challenges you will face is that of overcoming resistance to change.

If you find it necessary to overcome a certain amount of resistance, there are three steps you can take to help you effectively deal with this resistance:

- Determine the source of resistance
- Speak with those who seem to be resistant and attempt to find out what is behind this resistance
- Work to overcome the resistance

Chapter 1 discusses this concept of overcoming resistance to change in more detail.

Show that it will cost too much not to implement

As the individual who has chosen to champion the implementation of documentation testing in your company, you are likely to be required to prove that not implementing documentation testing will eventually be more costly to the company than it will be to implement it. When that time comes, you must be prepared to show that it will cost too much not to implement documentation testing (Fig. 7-4).

To accomplish that task you have to begin recording, reviewing, and analyzing information that shows the costs of continuing to do

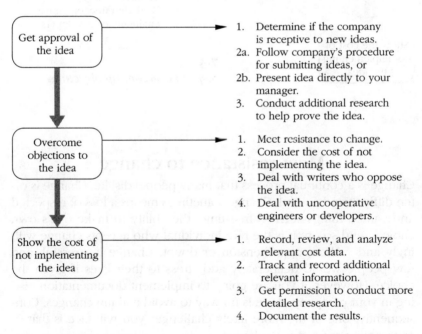

Get approval of the idea	1. Determine if the company is receptive to new ideas.
	2a. Follow company's procedure for submitting ideas, or
	2b. Present idea directly to your manager.
	3. Conduct additional research to help prove the idea.

Overcome objections to the idea	1. Meet resistance to change.
	2. Consider the cost of not implementing the idea.
	3. Deal with writers who oppose the idea.
	4. Deal with uncooperative engineers or developers.

Show the cost of not implementing the idea	1. Record, review, and analyze relevant cost data.
	2. Track and record additional relevant information.
	3. Get permission to conduct more detailed research.
	4. Document the results.

7-4 *Additional problems to overcome*

business the way you do it now—without documentation testing. The best place to begin that process is right in your own backyard. Begin collecting information about the cost of product documentation errors, inconsistencies, inadequacies, and other related problems. Start a log of your own or have others start logs that track any and all problems related to the company's product documentation. Every time you or other individuals use the company's product documentation, keep a record or a log of aspects related to using that documentation. For example, the log might include information such as

- What question you were trying to answer or what problem you were trying to solve when you turned to the documentation for help
- How long it took you to find what you were looking for
- How many places in the documentation you searched for the needed information before you found it
- How you finally found the information you were looking for (index reference, table of contents, the document's author told you where to look, and so on)
- Whether the information you found was complete (provided enough information at a sufficient level to solve your problem or answer your question)

Track any information relevant to the time associated with answering a question or solving a problem. Also track any additional information you think might be relevant, even though it is not listed here. Later, the information contained in these logs can be reviewed and summarized to provide an analysis of how much time is wasted because information is difficult to find, unavailable, incorrect, or otherwise insufficient. You can then multiply that time by an average hourly rate of pay for those involved in the tracking task to give a rough estimate of the costs associated with not making an effort to improve the documentation's quality.

You can take this effort one step further, moving out from your own experiences to record and track similar user experiences. If you have any business relationships with individuals or companies who use your company's product, you might be able to get one or more of them to participate in your research. Product users might be willing to keep the same type of log or record you were keeping, for a limited period of time. After that time, collect their records, review and analyze the information their logs contain, then associate costs to that lost time.

One way to associate the cost to your users is to estimate the customer's cost of time spent using your product's documentation. Calculate that cost in the same way you calculated it for members of your own company, by multiplying an average hourly pay rate by the number of hours spent using your company's documentation to solve problems or find answers.

The level of frustration and anger your own company employees feel as the result of working with the company's product documentation is important, and should be tracked on the logs as well. It is not as important to track it for company employees as it is to track it for those customers who are assisting you in collecting this information. Whenever possible, ask the customers to also record their reactions, emotions, and so on. If they get angry when trying to find information, ask them to state that fact. If they wish they could throw your product out the window, ask them to state that type of fact as well. Later, when you prepare your report to management, including a couple of appropriate quotes from users can help dissuade an otherwise undaunted individual into giving serious consideration to implementing documentation testing, at least on a trial basis.

If you have permission to conduct more extensive research, you might consider another option—that of going on-site to visit customers and see how they use your documentation. Intuit, Inc., the manufacturer of Quicken (personal financial software) and QuickBooks (accounting software aimed at small businesses) uses a similar practice to assess the success or failure of its products (Fenn 1994).

> *That's why . . . Intuit Inc. . . . actually has market researchers go home with customers and watch them install and use its software. . . . "The purpose is to watch people use the product in their own setting," says Suzanne Taylor, whose title is senior customer-insight manager. "It's like being a fly on the wall. It's a good way for us to learn from customers.*
>
> *"We observe how they experience the product the first time," says Taylor. Subjects are asked to "think out loud" and to perform specific functions, such as opening a file or entering data. Market researchers might, for example, look for changes in body language or notice if a customer seems puzzled by a certain screen. If someone has one or two simple questions during this process, the researcher helps; beyond that, they are referred to Intuit's technical support staff.*

While visiting the businesses or homes of several customers might not be a practical method of data collection for you, you still can take advantage of customer help. Questionnaires are a well-used method for collecting customer data, as are phone calls. In addition, you should consider bringing one or a few customers into your office and conducting a usability test with them. Establish data-finding requirements and procedures, then watch to see if they are successful at accomplishing them. This approach might cost you a little in time and "thank-you gifts" (t-shirts, mugs, lunch, and so on). It is, however, a proven and valuable method of collecting research information.

If none of these customer-centered options are available to you, go again to your own backyard for customer information. Contact your company's service and support or help lines. Find out if they keep any problem logs in which they track reported documentation problems. Ask to see the logs. Sort out the problems related to the documentation and create your own log of these reports. Include the length of time the service and support or help desk personnel spent attempting to solve the problem or find the information for the customer, then multiply the time by an average hourly pay rate for these personnel. You can then use the information you gather from this approach to help prove that not implementing documentation testing in your company is not only limiting the quality of your company's product documentation, but also resulting in added product support costs.

There are many approaches you can take to help support the idea that your company is spending money to make up for poor documentation quality. Your goal is to use one or more of these approaches to estimate how much money the company is spending, and then document that information as part of a management proposal to implement documentation testing. You can use the information you gathered about the cost of not utilizing documentation testing in your company, to prove that it is costing your company money by not implementing documentation testing.

Work with writers who do not want anyone to review their work

Once you get to the point where documentation testing has been approved for implementation, at least on a trial basis, you are likely to experience your next major problem. You might have to make great efforts to overcome the writer's dislike or fear of having someone review his documentation for the sole purpose of finding its flaws. Of course, while that might be the way the writer sees it, you should know that the primary purpose of documentation testing is not to pick on the writer. The primary purpose of documentation testing is to work with the writer to produce the best possible documentation given the time and other constraints under which that documentation is created. Now it is your job to convince the company's writers of the benefits of documentation testing.

To be fair, I should point out here that only rarely have I come across writers who did not want anyone to help them improve their documentation. Most writers willingly accept and often seek assistance from editors and peers in an effort to improve the quality of their work. When you do have a writer who is resistant to the idea of documentation testing, you must prove to that writer that

- Documentation testing is a quality-improvement process, not an attack on the writer. Sometimes writers have no idea what documentation testing is and thus are frightened of it, or they see it as a threat to them or their job. You must show them that documentation testing is simply one method for improving the overall quality of the product documentation, and not an attempt to destroy the writer's credibility, belittle their talents, or "get something on them" in order to get them fired or laid off.

- Documentation testing is a team effort that provides several benefits. Writers might find it difficult to see documentation testing as a way to make their job easier than as a way to increase their burden. If the writer believes that documentation testing will only increase their workload, you must convince them otherwise. You must help them see that, while there will be changes to the document that they will need to implement, the benefits you bring to documentation development offset the additional effort of making changes. Explain to them how you will be helping them with the development of that document as it progresses, as well as reviewing and correcting errors as early as possible to prevent serious problems (and extra work) later on. Review with them all of the benefits of documentation testing this book has previously explained.

- For the most part, documentation test results are only between the tester and the writer, and are not part of a plot to eliminate or discredit the writer. Writers sometimes believe that documentation testing results might be used to prove that the writer is incompetent. In most instances, the opposite is true. The documentation tester can point out areas in the documentation where excellent skill and knowledge have helped to develop a useful and cost-effective document. In addition, if testers look for the good areas of the document, as well as those areas that need improvement, a

documentation test can end up being a pat on the back to a
good writer.

- Documentation testers can actually be a resource instead of a
  burden to the writer. Writers sometimes fear that
  documentation testers will get in their way, interfering with
  their work. It is up to you to show the writer how you can
  help them instead of hinder them. For example, you might
  explain how you can answer questions for them when the
  product's developers cannot spare the time, or how you
  might be able to carry the responsibility of finding the answer
  they need for them, freeing them to go back to the process of
  writing. After all, much of your time as a documentation
  tester is spent with the development team during the
  product's development. You have relatively easy access to
  developers, and you are somewhat free to pursue and follow
  up on information gathering. At this particular stage in the
  document's development, you will not be bogged down with
  testing. You will have the time and the incentive to assist the
  writer with problem-solving, information-gathering treks such
  as these. As a result, the writer will have more free time to
  write, and you will know more about the product than you
  would have if you had not researched the answer to
  particular questions.

Figure 7-5 summarizes these points.

Regardless of how you approach the task of swaying a writer to
the belief you have in the effectiveness and need for documentation
testing, remember the writer needs to believe in it too. Therefore,
consider the writer's point of view when determining how best to
help the writer see documentation testing's benefits. Emphasize to the
writer the particular benefits that make the most difference to him.
Help the writer see where documentation testing can help, not hinder, him in his quest for quality documentation. Prove to him that the
documentation will be better for your testing efforts. You might also
want to remind him that he will be getting the credit for a job well
done. After all, it will most likely be his name on the final document,
not the name of the documentation tester.

Help engineers to cooperate with you

Once you have convinced the writer that having documentation testing
in the company is to his benefit, the next group of individuals you have
to convince is the product's developers or engineers. As with most
company personnel, engineers or product developers are pushed to

Writer's benefits of documentation testing

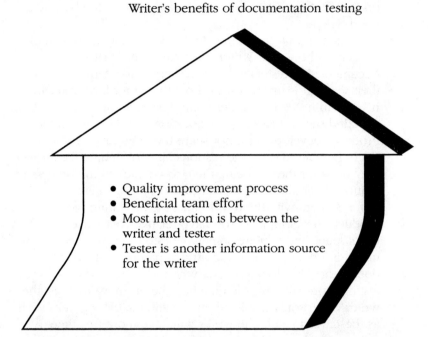

- Quality improvement process
- Beneficial team effort
- Most interaction is between the writer and tester
- Tester is another information source for the writer

7-5 *How to overcome a writer's resistance*

meet tight deadlines with limited resources. One of these resources is, of course, the amount of time each developer has to give to the design and development of the product.

Engineers might be reluctant to endorse or accept the addition of documentation testing and documentation testers to the development process if they believe it will cut into their currently available resources. In other words, they might believe documentation testing, and consequently, documentation testers, to be more of a burden than a blessing. When this is the case, you must work to dispel this belief and prove the merits of documentation testing to the engineer. In other words, you must convince the product's engineers to cooperate with the documentation tester (Fig. 7-6).

The first step toward accomplishing this goal is to convince the development manager that you should and need to be part of the development team. You must introduce yourself to the development team manager, explain your position as a documentation tester, then explain the purpose and benefits of documentation testing. You also need to show the development team manager how and where documentation testing fits into the development cycle. In addition, you must prove to the development team manager that documentation

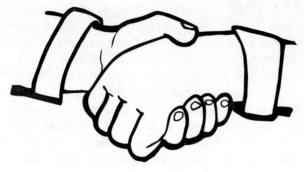

To help engineers to cooperate:

☐ Convince the development team manager of the value a documentation tester adds.

☐ Convince development team members of the value a documentation tester adds.

☐ Hold one-on-one meetings with development team members to discuss the benefits of documentation testing.

7-6 *Help engineers to be cooperative*

testing can fit successfully into the development cycle without adding to or distorting the cycle in any way.

With that out of the way, and your acceptance as a team member under way, you will most likely have to repeat the process for the development team. At the first meeting, it is probable that you will be asked to explain who you are, why you joined the team, what your role in the team is, how your role will benefit the product as well as the documentation and the other team members, how the engineers can help you to successfully meet your job obligations, and so on. Therefore, I suggest that when you attend that first development team meeting, go prepared with a concise and well-rehearsed minispeech. First impressions count, and this might be not only the first chance you get to convince the team to give you and documentation testing a chance, but it might be the only chance you get.

If the opportunity presents itself, there is one other approach you can take to winning the product engineers over to cooperate with you.

You can hold informal one-on-one meetings with them. This usually works best if you already know most or all of the product development team members. You should call them and ask if you might come by to speak with them, or you might just drop by to see if they have a moment of free time to chat with you. If you pass them in the hall or see them in the cafeteria, you might take that opportunity to mention your new association with the team and the responsibilities it carries with it.

Help them to understand what you do and why it is important. If you help them to see you and documentation testing as a method of helping them to create a better product, you are more likely to get a positive response than a negative one.

Develop a management proposal

To prove to your company's management that they should consider documentation testing, you might need to make only an oral presentation or request, with little to accompany that presentation or request. On the other hand, you might be expected to justify your request just to research documentation testing as an option, before they will even consider discussing it. The bottom line is that you might have to develop and present a formal management proposal, or you might choose to do so in order to help ensure the success of such a proposal.

If you have to prepare a formal proposal, there are many books and articles you will find useful to help you create a professional and effective proposal. In order to get management to make a decision to pursue documentation testing for your company, your proposal needs to show management two important items of information:

- How documentation testing fits with the company's mission
- What benefits the company can expect to receive from implementing documentation testing

Documentation testing and the company mission

All companies have a mission. Not all companies shout it to their employees, reiterate it at each opportunity, note it on posters and in employee publications, or in other ways directly point it out to their employees, but they have a mission nonetheless. It helps if you have a clear definition of what your company's mission is. Then, when you begin to develop a management proposal to suggest or recommend documentation testing, you can make sure that your recommendation is defined in terms of the company's mission.

For example, Novell, Inc., a leading manufacturer of networking software, has had as its mission statement that of "Growing the networking industry." To the mind of Novell's management, growing or enlarging the networking industry is beneficial not only to Novell, but to all interested and related parties. Novell put most of its corporate emphasis on efforts toward expanding the networking market. As a result, the more the networking market expanded, the more sales of networking software Novell made. Growing the networking industry grew Novell.

Novell already utilizes documentation testing as a standard practice of documentation quality control. If they did not, however, a Novell employee who wanted to convince Novell's management to implement documentation testing would have to show Novell's management how documentation testing fits into their overall corporate mission—to grow the networking industry.

To do this, the employee could concentrate on one of the most important benefits of documentation testing—improved document quality. However, the side benefit of improved product quality might be equally important to the company's corporate mission.

As noted earlier, one research study showed that when conducting documentation tests against documentation produced for a software product (specifically networking software), several errors were found in the software itself. In one instance, approximately 50% of those errors were not found by product testers performing a defined set of product tests. This early location and subsequent correction of product errors resulted in the company shipping a higher-quality product. The company firmly believed that, in order to grow the industry, consistently producing high-quality products was important. In this manner, documentation testing contributes to the company's mission.

When you develop a management proposal with the intent of selling the idea of implementing documentation testing in your company, remember the importance of the company's mission. Determine what your company's mission is. Then, to the best of your ability, show how documentation testing can help your company to fulfill its mission.

Identify the benefits

Showing how documentation testing can help fulfill the company's mission is often not enough to convince management to spend the time and effort developing a corporate policy on documentation testing. It might not even be enough to convince them of the need to research its use. You might also need to identify for management those benefits it can expect to receive as a result of implementing documentation testing.

Chapter 1 discussed several benefits of implementing documentation testing in your company. The benefits discussed were divided into long term and short term. The long-term benefits of implementing documentation testing in your company are

- Better document quality
- Better product quality
- More-usable products
- Fewer product releases
- Fewer document rewrites and errata

The short-term benefits of implementing documentation testing in your company are

- Reduced documentation page count
- Improved product design
- Reduced after-sale service costs
- Reduced training costs

Almost any company that produces documentation can benefit from implementing some form or level of documentation testing. Regardless of the type of product your company produces, if documentation accompanies that product, implementing documentation testing can help to reduce your long- and short-term costs of documentation development. As an added benefit, documentation testing might even help protect you from legal action should something in your documentation cause problems for the user.

To help convince your company's management that documentation testing is at least worth looking into, first take a look at your business and your company's mission. Make an effort to find ways in which documentation testing can be successfully implemented in your company. Look for ways in which specific aspects of documentation testing can bring your company the greatest return on its investment in documentation testing. Then, when presenting the idea to management, concentrate on the benefits documentation testing has to offer for your company.

Of course, you cannot ignore the drawbacks. You are likely to be questioned about the drawbacks, so be prepared to respond to these questions. Chapter 1 contained a discussion of the benefits as well as the drawbacks of documentation testing. Be prepared when you are questioned about documentation testing's potential overall effect on your company, positive or negative.

Get influential people on your side

As you get closer to presenting the idea of implementing documentation testing in your company, it might be in your best interest to test the waters and find out whether there is any support for your ideas. One way to do this is to introduce the idea gradually to individuals who have some say in the final decision.

To introduce the idea of documentation testing, you can take several approaches. For example, you can do as was suggested earlier—make a copy of this book available to your manager. Highlight the most important information in the book, then when you give it to your manager, ask for an opinion on the potential for documentation testing in your company.

Another approach you can take includes writing a brief but formal summary of documentation testing. You can use this book as a guide. Present it to people in your company who can both benefit from its implementation and help to ensure its successful integration into the company.

You can also meet with individuals and discuss the idea of documentation testing on a less-formal basis. When the opportunity presents itself, ask these individuals whether they have ever heard of documentation testing. If they have, find out what their definition of it is and what their experiences with it have been. If you are confronted by someone whose experiences were not positive, try to find out what the problem was. If and when you encounter resistance from this individual, you can deal with that resistance by researching ways in which to correct or prevent the problems and negative experiences the individual previously encountered.

However you approach the task, your goal is to plant the idea seed and nurture its growth. Developing positive support for your idea before you make the formal presentation increases the likelihood that you will be successful and have your idea accepted.

What to include in the proposal

Once you reach the point of being able to develop your proposal, there are some basic items of information you should include in your proposal (Fig. 7-7):

- Overview of the purpose and function of documentation testing
- Benefits of documentation testing
- Potential drawbacks associated with documentation testing

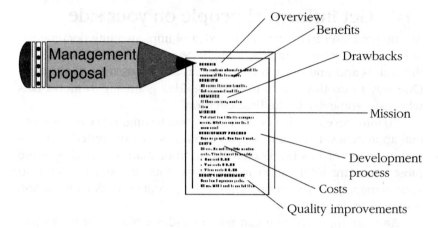

7-7 *Information to include in the proposal*

- How documentation testing will help the company achieve its mission
- How documentation testing fits into the company's current product and documentation development
- What it will cost the company if they do not implement documentation testing
- How the implementation of documentation testing can improve product and document quality, increase the company's presence in the market, save the company money, and any other factors that can improve the company's bottom line

Companies want to improve their products. They want to provide the best possible product for the lowest cost, and develop a successful customer relationship. Documentation testing can help a company do just that. It is not always easy for management to see how documentation testing can help improve their products, reduce their costs, and improve customer satisfaction. It is up to you to point out the possibilities that documentation testing provides a company in these areas. The best place to do this is in the proposal.

Make an effort to include at least these seven items of information in your proposal. Add other items of information which you believe might be important to your company's management. Then, prepare to present that proposal to your company's management and other interested individuals.

Present the proposal

More than one great idea has gone by the wayside because the idea's presentation was not all that it should have been. It is not always easy to determine what is important to your audience. Nonetheless, you must attempt to understand your audience and then develop and present a proposal your audience will find acceptable, or your idea, no matter how great, will stand little chance of being accepted.

Worse than having your idea rejected, however, is never presenting it at all (Griffith 1990).

Thinking is easy, acting is difficult, and to put one's thoughts into action is the most difficult thing in the world.
— Johann Wolfgang von Goethe

This book is designed to give you the information you need to understand and implement documentation testing in your company. You must then act on that information if it is to be of any use.

You might be in a position to make the decision to implement documentation testing, and then be in a position to act on that decision. If you are not, you must bring the possibility to the attention of those who are capable of making the decision. You must give them the information they need and the opportunity to use that information to the benefit of the company. Doing so is the second most difficult thing you have to do. Getting started is the first most difficult. Both you and the company can benefit from your actions, however (Griffith 1990).

One who gains strength by overcoming obstacles possesses the only strength which can overcome adversity.
— Albert Schweitzer

Whatever you do, if you believe documentation testing will benefit your company, develop and present your proposal to those who can act on it. To help you do that, this section discusses three areas of proposal presentation:
- The primary purpose of the proposal presentation
- Suggestions for creating an effective proposal
- Proposal presentation hints and tips

Primary purpose of the proposal presentation

It seems obvious that the primary purpose of presenting a proposal on documentation testing is to convince those in power to use that power to implement documentation testing. Sometimes all you can hope to do with your presentation is convince management to look into the idea of using documentation testing in the company. The real purpose of presenting the proposal, however, is to give life to that proposal, not just to sell the idea.

A written proposal can provide those who need to know with the information they need to know in order to make an appropriate decision. A written proposal, however, cannot carry with it the author's beliefs and feelings toward the idea conveyed by the proposal. Therefore, the real purpose of presenting a proposal is to bring those beliefs and feelings into the equation. Doing so helps to sell those in charge on the concept or idea. Once you have accomplished this, the idea and its implementation will help carry the process forward.

When you prepare to present your proposal, you want to create as effective a proposal as you can, and understand and utilize various hints and tips for making it the best proposal you can present.

Create an effective proposal

The primary ingredient in an effective proposal is the information it contains. That information must be accurate, complete, and of interest to the audience. In this case, the accuracy, completeness, and interest are assumed. Therefore, you need to concentrate on the physical creation and delivery of an effective proposal. That is, you need to be concerned with the successful presentation of your proposal. A checklist can help you to do this.

To develop a proposal for an effective presentation, keep the following points in mind (Sides 1991):

- The proposal must be clear and without errors.
- The proposal must be rigidly structured to make it easy for the audience to see how one point flows to another.
- The proposal must contain a limited number of the most important points.
- Each point covered in the proposal must be clearly identified and separate from each other point.
- The arrangement of presenting each point must be logical and orderly.
- Transitions from one point to the next should be logical and smooth.
- The proposal should contain a beginning, a middle, and an end.
- The proposal should be presented as effectively as possible.

There are many books and articles available to help you develop and customize your proposal. Do some research and reading to find the ideas and approaches you feel most comfortable with. Then write and prepare your proposal following those ideas and approaches. Once you have prepared your proposal, you then have to present it.

You might consider presenting your proposal in one of four ways (Sides 1991):

- Manuscript—providing the proposal in written format to be read to the audience in a formal setting such as a meeting.
- Memory—memorizing the entire presentation and then delivering it completely from memory.
- Impromptu—delivering the presentation on the spur of the moment without any preparation or forethought to doing so (not recommended).
- Extemporaneous—presenting the material from notes or an outline (the preferred method of presentation).

The preferred method of presentation for most speakers is the extemporaneous method. To make such a presentation, however, requires not only the development of an effective proposal, but also some skill at making presentations. The following section provides some presentation hints and tips to help you polish your presentation skills.

If your presentation skills are limited, consider taking a seminar or training course in presentations as a precursor to delivering your presentation on documentation testing. You do not, however, have to be a polished speaker to get your point across. If you remember and attempt to implement some of the hints and tips provided on the following pages, you should be able to make a reasonably effective presentation.

Presentation hints and tips

As with creating an effective proposal, there are many books and articles to help you improve your presentation skills. You might want to check out some of those materials or attend a class or training seminar. If time, money, or desire do not permit more detailed research and training in making presentations, try to implement some of the hints and tips given here to make your presentation as effective as possible.

When making a presentation, consider the following hints and tips (Mandel 1987):

- Take steps to reduce the anxiety and stress often associated with making a presentation. Some steps you can take include being organized and prepared for your presentation, imagine yourself giving the presentation, practice your presentation, breathe deeply, focus briefly on relaxing before you begin, exercise your muscles (tighten then loosen fingers, toes, and muscles to help relax them), move around, and keep eye contact with the audience, thinking of them as friends rather than enemies.

- Ensure your presentation is well prepared. You should be able to list the objectives of your presentation. Ensure that your presentation meets those objectives.

- Understand your audience, what their values, needs, constraints, knowledge, and hot points are so you can present toward those aspects of your audience.

- Plan and prepare appropriate handouts or visual aids. When doing so, concentrate on having those handouts focus attention on the main points you want to cover, reinforce the message you are trying to convey, stimulate the audience's interest, and help the audience to visualize concepts and ideas that otherwise might be difficult to perceive.

- Prepare for your presentation by practicing it. If possible, practice it in the room in which you will be giving it. Use the audiovisual equipment to understand how to use it so as to deliver your presentation smoothly. Find the best place to stand so that the audience can clearly see you and hear you, and so they can also see the visuals you have created for them. In general, make yourself comfortable with both the presentation and the surroundings in which you will give the presentation.

- When delivering the presentation, deliver it in the proper order. Begin with an introduction, followed by a brief statement about what the presentation is intended to convey. In other words, tell the audience what you are going to tell them. Next, present the body or main ideas of your topic. Present benefits related to your topic, documentation testing in this case, and any other relevant information. Next, summarize the presentation. That is, tell them what you have told them. Finally, answer any questions and close your presentation.

- Remember that your posture, gestures, and nervous habits can distract the audience from the actual presentation. Whenever possible, first give the presentation to a friend that will provide helpful feedback. Tell them you want them to watch for anything you do or say that distracts from the presentation. Thank them for their input and show your appreciation by taking advantage of the information they provide. Once you get that feedback, use it to improve your presentation.

Create an effective presentation, plan and prepare for it, practice its delivery, then try to relax when it is time to deliver it. The presentation you deliver is not likely to determine whether you continue as an employee with the company. It can, however, be an opportunity to advance within the company, maybe even advancing you into the position of documentation testing manager.

When creating and preparing to deliver your presentation, think of yourself as an entrepreneur. Consider this quote from Allan A. Kennedy (Griffith 1990):

> *To be an entrepreneur takes more than just a good idea. As the founder of Atari said, "the critical ingredient is getting off your butt and doing something. It's as simple as that. A lot of people have ideas, but [t]here are few who decide to do something about them now. Not tomorrow. Not next week. But today. The true entrepreneur is a doer, not a dreamer.*
>
> *"Entrepreneurs have an action orientation, but it is more than that. They want to build something, to make something. The pride is in taking a product and making it commercial."*

Take your *product*—the idea of implementing documentation testing in your company—and make it *commercial*. Convince your management that the entire company and the product's users will benefit from doing so.

Summary

Starting, developing, and managing a documentation testing group in your company can benefit everyone involved. Before you can do so, however, you have to overcome potential problems commonly encountered when proposing any new business idea.

The first step is to convince management that documentation testing should be a regular part of your company's development process. To accomplish this step you might have to conduct research, gather related information, and develop and present a management proposal. This chapter was designed to help you in three areas:

- Recognizing and overcoming potential problems you might encounter when trying to convince company personnel of the need for and benefits of documentation testing

- Developing a proposal for management to help sell them on the idea of implementing documentation testing
- Creating and presenting that proposal in the most effective manner possible

Just reading this book is not enough, however. As Allan A. Kennedy put it, "the critical ingredient is getting off your butt and doing something."

References

Fenn, Donna. 1994. "Street Smart." *Profit* (November/December 1994): 37–39.

Griffith, Joe. 1990. *Speaker's Library of Business Stories, Anecdotes and Humor*. Englewood Cliffs, New Jersey: Prentice Hall.

Mandel, Steve. 1987. *Effective Presentation Skills*. Los Altos, California: Crisp Publications, Inc.

Sides, Charles H. 1991. *How to Write & Present Technical Information*. Phoenix, Arizona: Oryx Press.

8

Hiring a documentation test team

Once you convince company management to start a documentation testing group in your company, you can begin the process of developing and staffing the group. You will have to consider such aspects of developing a group as preparing a budget and hiring personnel. This chapter is designed to help you get started in each of these aspects of establishing a documentation group within your company.

After reading this chapter, you will understand how to
- Develop a testing budget
- Establish employee skill levels and pay ranges
- Find and hire qualified people

Develop a testing budget

Staffing a documentation testing department requires finding and hiring qualified people. How many people should you hire? How much can and should you pay the people you hire? What benefits should you offer to them? All of these questions must be considered and answered based on company policy as well as one other important item. The answer to these and other related questions is determined, at least in part, by the amount of money you have been allocated to spend. That is, by the budget established for your department.

In some companies, budgets are developed by upper management. Middle management is then told what they have been allocated to spend. In other companies, budgets are developed by middle management, and are then approved by upper management. Regardless of the method used in your company, you might have the opportunity to contribute to the development of a documentation testing group's budget. If you do, you will need to have a basic idea of how to develop a budget, what factors should be included in it, and which resources you will need to include.

Having an understanding of the purpose of budgets and what a documentation testing budget should include is important when proposing the development of a documentation testing team. A realistic budget makes it easier for you to provide management with a clear picture of what it will cost to establish a documentation testing team. Even if your goal is simply to get your company to let you become a documentation tester on a trial basis, budgeting knowledge can be important. You might be expected to show your company what additional costs (above your salary) it can expect to encounter if it adopts documentation testing, even on a trial basis.

Budgeting is practically a science in itself. There are several ways to approach the budgeting process. There are many different books and articles available on the topic. Do some additional independent research to find the method best for you. Don't overlook your own company's managers as a resource.

Companies often have their own process to follow when developing and submitting an annual budget for approval. However, if you are not currently in a position to become familiar with your company's budget process, this section will help you get started. From this section, you will learn

- The purpose of a budget
- What a budget should include
- How to estimate resource requirements
- What a sample budget might look like

Purpose of a budget

The purpose of a budget is explained by its definition. The American Heritage Dictionary defines the word budget as

> *n. 1. a. An itemized summary of probable expenditures and income for a given period. b. A systematic plan for meeting expenses in a given period. c. The total sum of money allocated for a particular purpose or time period. 2. A stock or collection with definite limits.*

The main difference between the first definition of the word budget and the purpose of a documentation testing budget is that the documentation testing budget does not include estimates for income. The only exception would be if the entire purpose of your company was to conduct documentation tests on an independent basis. In place of income estimates, however, you might consider including estimates of the savings the company can expect to receive based on such factors as reduced service and support calls.

As an employee of a company considering adding documentation testing to improve the quality of product documentation, income estimates are not included. The company's entire annual budget includes income estimates, and except to the extent that your requests for a piece of that income pie are affected by the company's income, you will not have to be concerned about income estimates or projections.

You will, however, need to meet the first three specified definitions:

- An itemized summary of probable expenditures
- A systematic plan for meeting expenses
- The total sum of money allocated for a particular purpose or time period

The primary purpose of establishing a documentation testing budget is to provide an itemized summary of probable expenditures. If you are going to ask for money to spend on documentation testing, you need to specify how much you are likely to need. Providing an itemized summary of probable expenditures does just that.

A documentation testing budget is also a systematic plan for meeting expenses. It specifies what you will be spending the money on and how much you will be spending. If you break the budget down into expenditures by quarters (three-month increments), you can be even more specific in your spending plan, although this is not always necessary.

Of course, your budget should provide a total of all of the funds you expect to need. Your budget should provide totals for the sum of money allocated for particular purposes (documentation testing) as well as for a specific time period, generally one year.

What it should include

To provide a reasonable budget, you must include more accurate information than a guesstimate of the total amount of money you expect to need. You must specify what you will be spending the money on, and how much you will be spending on each item or group of items, thus providing an itemized summary of probable expenditures.

The categories for which you will have to provide budget estimates include

- Payroll
- Equipment
- Products
- Facilities
- Miscellaneous

Payroll

Payroll expenses include estimates for salaries, including those to be paid to any current employees the documentation testing team has, as well as salaries to be paid to any employees you will be hiring, and estimates of benefit costs. If the average amount of money your company spends for benefits per employee has already been established (the personnel or human resources department can usually help you with this type of information), use these figures. Otherwise, you will have to attempt to estimate what the costs of employee benefits will be.

To provide a reasonable estimate of payroll expenses, you will also need to specify the number of employees you anticipate hiring. This section of the budget should include all of the employees the documentation testing team will have when established.

Equipment

Next to salaries, one of the biggest expenses is the cost of equipment, although this will vary greatly from business to business. Your employees will need a computer, word processor, or typewriter of their own in order to prepare documentation testing reports. Whether they have the actual product to test will depend on the product. So too for whether you purchase the competitor's products for analysis and comparison.

For example, if you are documenting space shuttles, you probably will not have to include the purchase of a space shuttle as part of your equipment list. If you are documenting application software, while you might not have to purchase your company's software, you might have to purchase computers on which to run that software.

You might also need related equipment. For example, if you are testing a software application program, you might need a printer to test its printing functionality. You might also need a modem if the software provides modem connection, and so on.

Product

When the product is not part of the equipment list, or when you must purchase your own product, you need to include its cost in your budget. Although most companies provide their employees with the product against which they can conduct tests, some companies use a cross-department budgeting method. A company whose departments are required to track everything they produce is likely to "charge" other departments for any supplies (including their own product) that they provide. In most cases, you do not actually pay for the product. It is a paper transaction. This must be included in your budget, however, if your company operates on this basis.

You can also include under this category any product you need to actually purchase. This usually means you need an item in the budget to purchase competitor's products.

If your company is a publishing company specializing in business books, for example, you might need a budget amount set aside with which to buy books from other publishers. The purpose of purchasing competing products is to see how the competition does it, compare your quality and breadth of product to theirs, and perhaps to get ideas for other products. Regardless of the reason, and you should discuss this as part of your budget proposal later on, you will most likely need to budget for product purchases.

Facilities

You have to have a place for your documentation testers to sit and work. This might mean you need to budget for individual offices or working cubicles, as well as for a separate area in which to conduct tests (a documentation testing lab). You will need to furnish these areas with desks, chairs, conference tables, telephones, lights, storage, filing cabinets, paper, pens, paper clips, binders, and other office supplies.

Miscellaneous

Miscellaneous includes a variety of items. For example, if your company manufactures and sells sporting equipment, your documentation testers might need subscriptions to various sports magazines. They might also need to travel to conventions or sporting activities.

This category is a catchall for any item that does not fit into the other categories. When appropriate, and when your company has defined categories for budget items, put your budget items into those categories. Use the miscellaneous category only when what you are budgeting for does not fit any place else. The more items you put into the miscellaneous category, the more explanations you will have to provide when it comes time to justify budget requests.

In addition to the monetary items to be included in your budget, it is often necessary to provide justification for your expenses. For example, if you are asking for sufficient payroll budget to hire two new employees, you will be expected to justify why you need to hire two new employees. Most often, the best way to approach this justification is to include a list of the projects you expect to have the documentation testing team working on during the coming year. In addition, if your team will experience some lag time because of training requirements, or you anticipate it will be necessary for your documentation testers to conduct research on competitors' products, this information should be included in the budget as well.

Estimating resource requirements

In addition to asking for a specific amount of money for the documentation testing team, you will have to specify what your money will be used for. You can place dollar amounts into categories such as payroll, facilities, and so on, but you are likely to be required to provide estimates of the resources those sums will support.

Using payroll as an example, you will have to provide not only a dollar amount in your budget for payroll, but you will also have to explain how many employees that dollar amount is expected to cover. Further, you might have to break it down into a monthly amount. This is particularly true if you will not be hiring new employees on the first day of the year, but will instead be hiring them one per quarter or taking some other approach to staffing your documentation testing team.

If you are going to equip a documentation testing lab, you need to provide a detailed list of what you are going to include in that lab. You might be required to specify such details as how many telephones you will be placing in your lab, or the number of square feet of working area you will need for each documentation tester.

Budgeting is a process best started as early as possible. In some companies, the level of detail is not as critical as it is in other companies. Also, once the documentation testing team is established, the level of detail you need to include in your budget might be less than is initially required.

How you go about establishing a budget depends on what your company requires. The best way to determine this is to talk with others who are responsible for establishing budgets. If you can, get them to give you a copy of an old budget they prepared. If they are not comfortable with this, perhaps they will be willing to provide for you a copy of a budget that has no dollar amounts filled in. Otherwise, speak with the accounting (or similar) department, and ask them for help in preparing your first annual budget.

Establish skill levels and pay ranges

If you are going to establish a documentation testing team or department in your company, you must consider how experienced and educated your testers must be. In addition, you must set up a path for promotion, establishing levels through which your testers can pass as their knowledge and skills grow. This section is designed to help you accomplish these tasks. It provides you basic information from which to start.

Skill levels

Most jobs have various levels of skill for employees: beginning (or apprentice), intermediate, and advanced. There are often levels within these levels as well. For example, the advanced level might include a regular advanced level and an advisory or consultant level which would be the next step above the advanced level.

The three basic levels—beginning, intermediate, and advanced—are discussed here. Information about the level of skill and knowledge required for a documentation tester to qualify for each level is included.

Beginning

The functions of a documentation tester at the beginning level include the ability to evaluate documentation for technical accuracy, usability, and clarity, and the ability to support and participate in the development of a product and its documentation.

Minimum knowledge or experience and education requirements for the beginning level include

- A certificate in English, a related field, or in the discipline most closely associated with the company's field. For example, if the company develops software products, a certificate in computer science, robotics, electronics, or other related fields would be acceptable.
- Basic written and verbal communication skills.
- Basic knowledge of the technology associated with the product field. For example, if the company produces toasters, the beginning-level documentation tester should have an understanding of how toasters operate.
- The ability to work well in a team environment, as well as to get along with coworkers.

- The ability to deal professionally with stressful situations and to deliver under pressure.

The primary duties and responsibilities of a beginning-level documentation tester include

- Conducting documentation tests as assigned
- Participating and contributing as a tester
- Working on special projects as assigned
- Performing other duties as assigned

The beginning-level tester will probably rely on more senior testers and supervisors to determine the level of test to be conducted. Other more senior team members might also prepare the documentation testing plans and schedules, and confirm or make the relate commitments such as the level of test and start and complete dates.

Intermediate

The functions of a documentation tester at the intermediate level include the ability to evaluate documentation for technical accuracy, usability, and clarity, and the ability to support and participate in the development of a product and its documentation. The difference between the beginning and intermediate level is primarily the amount of support and training the beginning-level documentation tester needs, as opposed to the intermediate-level tester.

Minimum knowledge or experience and education requirements for the intermediate level include

- An associate degree in English, a related field, or in the discipline most closely associated with the company's field.
- Advanced written and verbal communication skills.
- Advanced knowledge of the technology associated with the product field.
- The ability to work well in a team environment, to get along with coworkers, and to mediate professional disagreements. For example, if two documentation testers are assigned to a project and they do not agree on the level of test to be conducted, the intermediate-level documentation tester should be capable of successfully mediating the disagreement to the benefit of the product documentation.
- The ability to deal professionally with stressful situations, and to deliver under pressure.
- The ability to assist beginning-level documentation testers, including providing training and peer review as needed.

The primary duties and responsibilities of the intermediate-level documentation tester include those assigned to a beginning-level tester, although the intermediate-level tester will be expected to fulfill those responsibilities and duties with limited or no assistance from others. In addition, the intermediate-level documentation tester might be expected to perform the following additional duties:

- Assist beginning-level testers to plan for, prepare for, and conduct documentation tests as assigned
- Work on more advanced special projects as assigned
- Mentor or train beginning level testers as needed
- Create and present documentation testing plans

Advanced

The functions of a documentation tester at the advanced level include all of those functions expected of the intermediate-level tester. In addition, the advanced-level tester is expected to train and in many instances ensure that other levels of testers successfully complete their assigned responsibilities. Advanced-level documentation testers might be given project management responsibilities, such as ensuring the entire project is successfully completed, even if that means bringing in additional testing help, rounding up equipment or products, settling disputes, and so on.

Minimum knowledge or experience and education requirements for the advanced level include

- A bachelor's degree in English, a related field, or in the discipline most closely associated with the company's field. For the advanced-level tester, a master's degree might be required in some of the more technical fields such as physics or biological sciences.
- Advanced written and verbal communication skills, as well as advanced editing skills.
- Advanced knowledge of the technology associated with the product field, and experience working with the product.
- The ability to work well in a team environment, to get along with coworkers, to mediate professional disagreements, to develop and schedule documentation tests, and to manage documentation testing projects.
- The ability to deal professionally with stressful situations and to deliver under pressure.
- The ability to mentor intermediate-level documentation testers, including conducting training sessions, testing reviews, and project review meetings.

The primary duties and responsibilities of the advanced-level documentation tester include those assigned to a intermediate-level tester, as well as the responsibility for managing products. In addition, advanced-level testers might be expected to review current testing procedures and policies, and provide and implement improvements to those procedures and policies. The advanced-level documentation tester might also be expected to perform some team-leader duties, additional mentoring or training, and to act in an advisor capacity assisting the development and documentation teams, as well as other documentation testers, to meet their goals and schedules.

Potential pay ranges

What you choose to pay your employees is a matter of company policy. It might also be dependent on the value your documentation testers add to the company and its products. This value can be used to justify the level of pay you set for your employees.

When establishing pay ranges, there are several factors you should take into consideration. Those factors are

- Pay ranges currently established for other related positions within the company. Your company's human resources or personnel department might have established job descriptions and pay ranges for each job within the company. If that is the case, you can discuss with them the types of responsibilities and educational or experience requirements your documentation tester positions require. They can then provide you with copies of position descriptions and pay ranges for job titles with similar requirements.

- The maximum dollar amounts allocated for the documentation testing team. If your budget is limited, so too must be the salary you can offer to documentation testers. If you are forced into a minimum budget, you must adjust how much you can offer. Another approach is to hire lower level (beginning or intermediate) testers.

- The going rate in this and related fields. You will find it difficult to obtain salary information for documentation testers. While there are publications providing this type of information for just about any job you can list, the title of documentation tester is not currently listed among them. You can look for salary ranges in related fields such as writers and product testers, however, as skill levels and even some of the responsibilities are similar.

    One way to determine the average salary paid to employees in different fields is to read current issues of business magazines such as *U.S. News & World Report*. The

October 31, 1994, issue contains an article titled "20 Hot Job Tracks." This article discusses 20 jobs in areas where demand is expected to grow over the coming years. It gives salary ranges for entry, mid-level, and top positions, and provides related information such as what training is required, where the best places are for this type of job, and what other related jobs are considered to be hot job tracks (Friedman et al. 1994).

- The pay ranges being offered by other companies in the same geographical area. Even if no other company within your geographical area has a position remotely similar to that of a documentation tester, you must still consider the pay ranges of other businesses. Most people work because they have bills to pay, not because they love to work. You have to compete for employees with the salary you offer, regardless of whether the employee can find a similar position. A potential employee who can accept a position as a documentation tester or one as a writer or instructor, for example, is likely to choose the writer or instructor position if it pays substantially more than does the documentation tester position.

- The cost of living in different areas. Some companies are very geographically diverse. They might offer the same types of jobs in different regions of the country, or even in different countries. The economy and cost of living in the local area must be considered when setting salary ranges.

  For example, a documentation tester working in New York City might require a salary that is several thousand dollars higher per year than another employee working in Spanish Fork, Utah, doing the same job for the same company, because the cost of living in Utah is typically lower than the cost of living in New York.

- The pay ranges considered typical for the same or related fields. To be successful at setting pay rates for documentation testers, you need some idea of what this and related industries already pay for this or similar jobs at varying levels. If you can get information about pay ranges from other companies you will be one step ahead. Many companies, however, keep their salary information very confidential, and will not give you detailed information. If you call and ask, you might be able to find a general range. Of course, some of the same publications that provide basic job title and description information also provide salary information. (Contact your local reference librarian for help in locating such publications.)

- The level of success you are having with hiring qualified people. No matter how competitive your salary ranges, how many benefits you offer that other companies do not, or how fun the job might be, if you don't get applications from qualified individuals, you might have to adjust your salary ranges. It might be necessary for you to offer a hiring bonus to attract employees, increase the standard salary ranges, or even reduce the position requirements to find sufficient applicants to succeed in your hiring efforts.

Advancement through the levels

Once you have your documentation testing positions established, you must make it possible for individuals to progress through the different levels. People want to be recognized and rewarded for their efforts. If you do not make it possible for your documentation testers to be rewarded with advancement opportunities, you will lose them to other positions or other companies.

Providing the opportunity for testers to advance in their positions is particularly important in the field of documentation testing. As you might have already figured out, it does not take long for your documentation testers to become some of the most valued employees in the company. Documentation testers soon gain more knowledge about the company's products than almost any other employee. They learn the products in intimate detail, function well in team settings, write effectively, and are self-starting and highly motivated individuals. Within two years, sometimes less, a documentation tester is generally ready to move up or on, and will find a wealth of opportunities outside your department if you do not provide them within your department.

There are two factors in assessing whether a documentation tester is ready to advance to the next level. They are longevity and skill.

Positions can be designed to qualify an employee to move to the next level based on their longevity with the company. When an employee has been in his current position for a certain length of time, which is defined in the job description as a minimum number of years of experience, he is ready for a promotion. In many instances, this is 2 to 5 years per level, but might be as long as 10 years. Given the previous statement that testers are often ready to move on after as little as two years, this approach is rarely successful.

The second approach—meeting minimum position requirements—is usually more effective. This means that in order to make it
possible for a documentation tester to move to the next level, you design positions and establish policies that let testers grow into the minimum requirements of the next level. Once the tester meets the
minimum qualifications for the next level, the tester can then be promoted. Of course, this assumes that the employee's personal habits
(arriving on time to work, keeping the company's regularly scheduled hours, and so on) are satisfactory. Assuming they are, you can
set up positions so that testers can be promoted as soon as they meet
the minimum requirements of the next level. If you take this approach, there are two things to keep in mind.

First, set a minimum length of time the tester must be in the current position before they are eligible for promotion. In many companies, six months to one year is typical. For a tester to move to the next
level after only six months, however, this tester must have been almost
overqualified for the level at which you hired him for to begin with.

If you set a minimum length of time, do not make it too long. If
you exceed two years, for example, you might find that testers are
unwilling to stick around that long to receive a promotion. Again, the
problem is that documentation testers tend to grow very quickly in
both their personal and professional skills and knowledge. You must
plan for such rapid growth or expect to hire and train replacement
testers on a regular basis.

Testers can soon become bored with their jobs if they are not
continually challenged. "An industrial psychologist stated that a number of problems in business and industry are a result of boredom.
This lack of job satisfaction is not caused by repetitive jobs but rather
by resentment and depression that workers feel when they are de-
prived of the opportunity to use their skills" (Griffith 1990).

Second, establish growth plans for each tester. Tell them what
they need to do to meet their current job responsibilities. But beyond
that, show them what the next level for their job requires. Help them
to plan and grow towards that level. They are likely to get there without your help, but if you define a clear path to the next level and give
them the opportunity to progress along that path, they are far more
likely to want to remain as a documentation tester.

Documentation testers know they have other opportunities. In
fact, opportunities will come looking for them. However, they will be
less inclined to seek out other opportunities, or even to accept those
presented to them, if they know they can grow in their current positions. After all, they probably became documentation testers because

they like the work and the pay. In addition, making major changes in your life is stressful. Make it possible for the testers to grow and be rewarded where they currently work and you are far more likely to keep them as long-term employees.

One other suggestion, if you really want to keep your testers for a long time, is to establish levels within levels. For example, make two or three intermediate levels through which the tester can progress before they progress into the advanced level. Such an approach can help you keep your testers for several years.

Find and hire qualified people

Now that you know what levels of positions you have to offer, how to find out what levels of pay those positions can qualify for, and what you can expect from your documentation testers, you must find and hire qualified people. Oh!, if it were only that simple, but it is not. If you want to find and hire the best qualified people for your documentation testing team, you must proceed through a logical process. That process requires that you

- Define your needs
- Write appropriate job descriptions
- Write effective job announcements/advertisements
- Screen initial applicants
- Conduct interviews
- Rank the applicants
- Choose the right person

Define your needs

One of the more difficult aspects of hiring employees is determining or defining what you need for an employee. The best approach to defining your needs is to first define what you want these employees to accomplish for you. In other words, you need to understand what you expect the person who fills the position to do in that position. After having read this book, you should have a pretty good idea of what a documentation tester is responsible for.

To define your needs you also need to understand your company. To help you do this you can prepare a single-page summary of the company. Include information about its product line, its purpose, and its location. This will not only focus your attention, but will give you information about the company that you can give to new or prospective employees (Fig. 8-1).

CarryPhone Company
Company Overview

Background Information

The CarryPhone Company was organized in April 1989.
CarryPhone manufactures and sells mobile telephone carrying,
locking, and tracking devices for use by companies with an
installed base of portable telephones.

Dr. Jim Donson started the company shortly after thieves broke
into his small business office and stole his employee's portable
telephones and other electronic devices. After being told by the
police officers investigating the theft that the chances of retrieving
the phones and clearly identifying them was less than 20
percent, Dr. Donson was determined to design a way to carry,
lock, and track stolen or misplaced portable telephones.

Dr. Donson's CarryPhone Company has grown from three
employees working out of Dr. Donson's five-employee business
office to the ten-story office building it now occupies in Carborn,
Mississippi. The company has three distribution warehouses; one
in the northeast, one in the midwest, and one in the west.

PRODUCT: The CarryPhone Security System into which you
place your portable phone, activating the
identification and alarm system.

SERVICE: Tracking systems are established in the three
warehouses. Purchasers report missing or stolen
portable phones and the satellite-based tracking
system installed in the three CarryPhone
warehouses tracks and locates the security
carrier's signal.

8-1 *Sample company summary*

Write appropriate job descriptions

Once you have defined your needs, you should write the appropriate
job descriptions. One approach to writing effective job descriptions is
to write them so they are results oriented. That is, write them so their
concentration is on the essential results you expect from an individual working in that position.

For example, some of the essential job results you might include
in a job description for a test technician (a results-oriented job de-

scription that is similar to that of a documentation tester) might include (Plachy and Plachy 1993)

- Prepares tests
- Identifies product capability and reliability
- Resolves testing problems
- Documents and reports test outcomes
- Maintains testing environment and information database
- Contributes to team effort

When preparing documentation tester job descriptions, you also need to write a job description for each level of documentation tester you plan to hire. Initially, you might want to hire only experienced testers. If that is the case, you should write that job description first (Fig. 8-2). You can write additional job descriptions for the beginning and intermediate levels later.

A job description should contain several basic items of information:

- A brief description of the position. This should include a description of the position's primary responsibility, as well as its most important additional responsibilities. For example, the primary responsibility might be written in a manner similar to the following:
  Reviews, evaluates, and tests end-user documentation to ensure its technical accuracy, usability, completeness, and clarity.
- Information about to whom this position reports. For example, a statement similar to the following might be included:
  The position reports to the Manager of Documentation Testing.
- An overview of the required education and skills. The job description should include a description or list of the minimum education and training required to qualify for the position. For example, a bullet list of education and skills such as that shown in Fig. 8-3 can be included.
- Information about the primary duties and responsibilities of the job. As with the overview of the required education and skills, the primary duties and responsibilities can be listed in bullet list (Fig. 8-4) or paragraph form. For example, a paragraph description of the principal duties and responsibilities of an advanced-level documentation tester might be written in a manner similar to the following:
  The primary duties and responsibilities of the Advanced Documentation Tester include planning, preparing for,

Job Description
CarryPhone Company

Position Title: Advanced Documentation Tester
Immediate Report: Documentation Testing Manager
Department: Product Development
Date: 29 June 1994

POSITION DESCRIPTION

The advanced documentation tester is responsible for testing product documentation to ensure its technical accuracy, clarity, usability, and completeness. Little direct supervision should be required to accomplish this testing. In addition, this position requires the employee to mentor and assist less-experienced documentation testers, and to fulfill all primary jobs functions and any other duties as assigned.

EDUCATION AND SKILLS

o Bachelors degree in English, electronics, or a related field
o Five or more years experience testing product documentation
o Experience with cellular phone or related technology
o Ability to function effectively with limited supervision
o Experience in user interface design and related skills
o Ability to work well with others
o Ability to meet deadlines
o Experience in training other employees

PRIMARY JOB RESPONSIBILITIES

o Plan, prepare for, and conduct thorough tests of product documentation
o Manage documentation testing projects as assigned
o Mentor and teach less-experienced documentation testers
o Provide team management skills as assigned
o Assist in the development of product documentation
o Actively participate on product development teams
o Work to improve the processes and procedures of documentation testing

ADDITIONAL JOB RESPONSIBILITIES

o Ensure internal and external customer satisfaction
o Promote the development and efficiency of documentation testing both within the company and within the field
o Continue to develop technical knowledge and skills
o Work on special projects as assigned
o Perform other duties as assigned

8-2 *Sample documentation tester job description*

Education and Skills

- Bachelor's degree in English, electronics, or a related field
- Five of more years experience testing product documentation
- Experience with cellular phone or related technology
- Ability to function effectively with limited supervision
- Experience in user interface design and related skills
- Ability to work well with others
- Ability to meet deadlines
- Experience in training other employees

8-3 *Sample list of education and skills*

Primary Job Responsibilities

- Plan, prepare for, and conduct thorough tests of product documentation
- Manage documentation testing projects as assigned
- Mentor and teach less-experienced documentation testers
- Provide team management skills as assigned
- Assist in the development of product documentation
- Work to improve processes and procedures of documentation testing

8-4 *Sample list of primary duties and responsibilities*

and conducting thorough tests of product documentation. In addition, the Advanced Documentation Tester is responsible for managing projects as assigned; mentoring, teaching, and leading less-experienced documentation testers; and assisting in the development of product documentation.

- Information about any additional responsibilities of the position. In addition to the primary duties of a documentation tester, there are generally other duties the tester must perform, and which, therefore, should be included in the job description. Some examples include ensuring customer satisfaction, promoting the development and efficiency of documentation testing within the company, working on special assignments, and other duties as assigned.

Once you have written a clear and concise job description, you can move on to the next step in finding and hiring documentation testers, that of writing effective job announcements and advertisements.

Write effective job announcements/advertisements

It might not be your responsibility to worry about advertising for potential employees. You might have a human resources or other department at your company that takes care of these hiring details. You might only be responsible for providing a position job description. After that, someone else has the responsibility of bringing resumés and applications from qualified individuals into the company.

If you do have to write the advertisement or job announcement yourself, try implementing one or more of the following suggestions to help you prepare an effective announcement or advertisement:

- Include a brief list of the most important skills, generally just the primary responsibility is sufficient.
- Include the minimum educational requirement. If you require a bachelor's degree, indicate that as a requirement. Include the field(s) in which you consider a bachelor's degree to be acceptable.
- If you are willing to accept an equivalent amount of experience in place of a degree, state so. For example, you might state that "the position requires a bachelor's degree in English, computer science, or a related field, or equivalent experience."
- Begin the advertisement with a catchword or phrase. Often this is simply the position title. It can also be a reference to the purpose or function of the job. For example, you might choose "Ensure quality documentation" as the lead of your advertisement.

If you still find yourself having difficulty developing a position announcement or advertisement, there are many sources of assistance available to you including your local librarian, other position announcements and advertisements your company has successfully used, and so on. You can also review the classified advertisements in the publication in which you will place your position announcement or advertisement and see which ones catch your eye. Then create something similar for the position you are advertising.

Screen initial applicants

If your position announcement or advertisement is successful, you will soon begin receiving applications and resumés. You must now screen these documents and choose the most likely candidates.

To screen initial applicants, compare their resumés or applications against the job description. Look to see which applicants meet the minimum education and experience requirements. Separate those applicants from the others. Those who meet the minimum requirements should be contacted to set up an appointment for an interview.

Conduct interviews

Bring the potential applicants into your company and interview each one. One of the goals you want to achieve during these interviews is to narrow down the list of candidates to a small number, preferably only two or three. From that list you need to make your selection.

There are many good references that will help you conduct successful interviews. There are also training seminars you can attend, video series you can view, and other sources of information available to help you learn to conduct good interviews.

If you are not experienced at conducting interviews in today's marketplace, you should seek out the information you need. Conducting successful interviews that are legal and effective is almost a science (or an art) in itself. In today's business environment, it is important to be as efficient and effective as possible. In addition, it is very important to ensure that you conduct your interviews and make your employee choices within the law.

Rank the applicants

Once you have chosen those applicants most likely to be successful in the advertised position, you must rank them. Rank applicants by their qualifications, and only by their qualifications. You need to consider whether or not their personality fits in with the company, if their personal habits are satisfactory, and related issues. You cannot, however, consider such things as their gender, race, color, national origin, and other factors. It is important for you to know what you can and cannot include when ranking the applicants. To be safe, consider only the applicant's qualifications, answers to appropriate questions asked during the interview, and other information the applicants provide.

One effective method of ranking applicants is to apply a point system to both the value of the questions asked during the interview and the applicant's response to those questions, and to the education and experience the applicant has. To keep it simple, use a 1-2-3 approach. Assign 3 to the most important factors or highest successful answer, and 1 to the least important or least successful answer.

Tabulate the applicant's scores. First, multiply the value you assigned to the question or experience, by the applicant's score. Do this for each question or experience. Add all scores to get a total score. Do this for each applicant. Then rank the applicants from highest to lowest based on their total score.

For example, if you believe that a bachelor's degree is of the utmost importance, rank it as a 3. Then, for each applicant who has a bachelor's degree, rank their type of bachelor's degree against the type of degree you consider to be most important. If your company produces software, you might give more weight to a bachelor's degree in computer science (ranking it as a 3) and less to a general business degree (ranking it a 2). Multiply the first factor—value of the degree (3)—by the second factor to get a total. In this example, for the applicant who has a bachelor's degree in computer science, the score would be a 9 (3 × 3). For the applicant who has a bachelor's degree in business, the score would be a 6 (3 × 2).

Choose the right person

Once you rank the applicants, it should be easy to choose the best candidate. Simply choose the candidate with the highest score. Unfortunately, it is not always that simple. A matter of just a few points difference between applicants, less than 10 for example, can make it difficult to choose. You must do so, however.

If necessary, you can have the final candidates interview briefly with other team members or management staff and let them give you their feedback. Document their feedback, then make your decision. If it is still too difficult to decide between two or three excellent candidates, consider asking your personnel or human resources department to help you make the decision. Whatever you decide, you must choose a final candidate, then offer them the position.

It is important to make the offer in writing. Once the offer is accepted, notify the other candidates in writing that someone else has been chosen, but that you will keep their application on file. After all, if the choice was really difficult, the other candidate might be a good choice for the next available position, saving you much time and money.

Summary

In order to establish a documentation testing department in your company, there are three major tasks you must complete. First is that of developing a testing budget. Second is establishing employee skill levels and pay ranges. Third is finding and hiring qualified people. None of these tasks are quick or easy. Much effort goes into creating a successful documentation testing team. In the long run, however, I believe you will find the effort to be well worth the work.

References

American Heritage Dictionary, 2nd college ed., s.v. "budget."

Friedman, Dorian, Dana Hawkins, D. W. Miller, and Andrea R. Wright, compiler. 1994. "20 Hot Job Tracks," *U.S. News & World Report* (October 31, 1994): 110.

Griffith, Joe. 1990. *Speaker's Library of Business Stories, Anecdotes and Humor.* Englewood Cliffs, New Jersey: Prentice Hall.

Plachy, Roger J., and Sandra J. Plachy. 1993. *Results-Oriented Job Descriptions.* New York: American Management Association.

9

Training and motivating a documentation test team

In an ideal world, every employee you hire as a documentation tester would have experience as a documentation tester and an intimate knowledge of your product, its market, and its users. Those employees would also know all there is to know about your product's competition, and would be skilled writers, editors, and speakers. Those employees would also be experts in user interface design, product development, and documentation development. And they would be willing to work long hours for reasonable pay. In short, those employees would be perfect.

If you can find and hire anyone who fits this description, good for you. However, it is far more likely you will hire employees who need additional training and experience in one or more of these areas. Ensuring they get the additional training and experience they need is as much the responsibility of the employer as it is the employee. The employee needs the desire and the willingness to learn and grow. The employer needs to provide the opportunity, and many times the resources as well.

A company can spend a great deal of time and resources growing each employee to be the best they can be. Once the employee feels he has learned everything he can in his current position, he will be ready to move on. If the company does not continue to provide the employee with growth opportunities, as well as motivation, the employee will move on.

There comes a point at which the employee cannot measurably grow in his job as a documentation tester. No matter what you do, you cannot prevent the employee from reaching his peak. At this point, the employer has two options. He can help the employee grow and move into another position in the company, or he can let the employee move elsewhere.

Obviously, the employee's move costs the company either way. It costs the company less, however, if the employee moves into another position in the company than if he leaves the company.

If you want to have the best quality employees and keep them as long as possible, you can help to do so by properly training and motivating them. Doing so ensures they continue to grow both personally and professionally and helps keep them motivated to remain with the company.

This chapter is designed to assist you in accomplishing two major tasks. The first task is that of defining and establishing a policy and process to meet employee training needs. The second task is to develop ways to motivate your employees so they want to be successful and continue working for your company.

After reading this chapter, you will understand approaches you can use to

- Satisfy employee training needs
- Motivate the documentation testing team

Satisfying training needs

No matter what the problem, you cannot solve it until you understand what is causing it. The same applies to training. You cannot define adequate and successful training plans until you know what employee training needs have to be met. The first step in satisfying employee training needs, therefore, is to assess your employees' present skills. Once you have accomplished that task, you can proceed with completing the other necessary tasks (Fig. 9-1).

Develop a skills assessment checklist

The first step toward developing your employees' skills and knowledge is to assess their present level of skill and knowledge. In making this assessment, the approach you take is determined by what you need to accomplish. In the case of documentation testers, what you need to accomplish is to help them be the best possible documentation testers they can. You therefore want to assess any skills, education, and knowledge they have that helps them to be a good

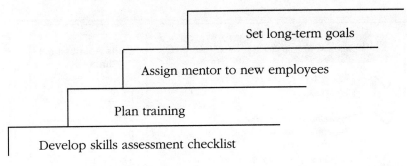

9-1 *Steps to satisfy training needs*

tester. By looking at what they currently know, you can then determine what they do not know. That determination gives you a starting point for establishing a training plan.

The first step toward assessing what your documentation testing employees already know related to documentation testing is to make a checklist of what they should know, called an *assessment checklist*. Each employee's present skills and knowledge are compared to that list. You can then create a training and development plan consistent with the company's and employees' needs.

Developing an assessment checklist is the most important part of properly assessing your employees' training needs. You should, therefore, give the development of this checklist as much time and attention as is necessary.

There are different approaches you can take to developing an assessment checklist (Fig. 9-2). For example, you can ask existing documentation testers, as well as people who work with them, what skills they believe are necessary for a documentation tester to be successful, and create a checklist based on their answers. Or you can assign a more senior (advanced) tester to create a list of needed skills and education. Either of these approaches might be sufficient, but there is a better way.

Perhaps the most logical, methodical, and complete way of creating a skills assessment checklist is to develop it against the documentation tester job descriptions. This method is the approach discussed here.

Before you begin developing a skills assessment checklist using this approach, you should be aware that there is one drawback. The assessment checklist will only be as effective as the job description is accurate and thorough. It is additionally important then that the job descriptions be well written and thorough to begin with. If you do

Developing a Skills Assessment Checklist

✓ Ask testers, peers, and others for their input.

✓ Have the human resources department create the checklist.

✓ Assign an experienced tester to create the checklist.

✓ Copy, with permission, a checklist used by another company.

✓ **Develop the checklist against the job description.**

Skills Assessment Checklist

☐ _____
☐ _____
☐ _____
☐ _____
☐ _____
☐ _____
☐ _____

9-2 *Approaches to developing a skills assessment checklist*

not believe this to be the case, you must consider rewriting and updating the job descriptions, or attempt to account for their weaknesses when you create the skills assessment checklist.

Each of the steps to follow when developing a skills assessment checklist are shown in Fig. 9-3. Each step is also discussed in this chapter.

To begin developing a skills assessment checklist, start by reviewing the job descriptions for each level of documentation tester. This helps to ensure that you include everything of value on your checklist. In addition, it helps to ensure you develop training plans that will help each employee grow into and prepare for the next level of promotion open to them.

When you review each job description, begin by looking for common categories of job requirements (Fig. 9-4). Some categories will be easier to define than others. You must try to account for all of them. The more-experienced documentation testers will already meet those categories found on the job descriptions related to less-experienced testers. Therefore, when looking for common categories of job requirements, begin with those that are most common, adding those that are not necessarily common to all levels as you go.

When looking for common categories, look for categories affecting all documentation testers, regardless of how experienced they are. All documentation testers, for example, will be responsible for

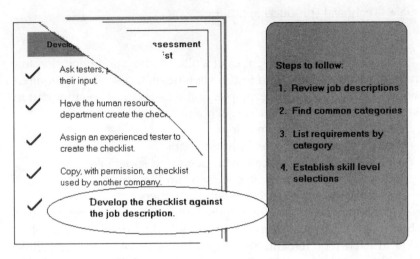

9-3 *Steps to creating a skills assessment checklist*

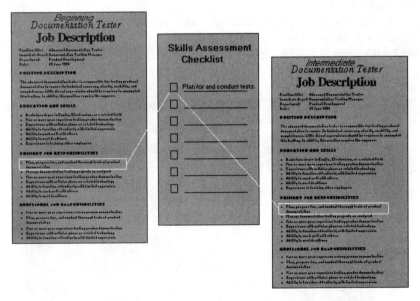

9-4 *Sample of reviewing job descriptions for common categories*

conducting documentation tests. This is a category of the job's requirements. Other categories likely to be found in job descriptions include

- Communication skills
- Professional behavior

- Technical (product) aptitude
- Education
- Team skills

First, define those categories of skills, training, and education important to a documentation tester. Then begin to list what is required within each category. For example, in the category of communication skills the requirements include such abilities as

- Ability to speak in meetings
- Ability to make formal presentations
- Ability to communicate with peers and superiors
- Ability to write clearly, concisely, and legibly
- Ability to edit materials written by others

Once you have listed the requirements within each category, you then make it possible to assess each individual employee's level of skill in each requirement. This might be as simple as placing three checkboxes next to each requirement (Fig. 9-5). Those three checkboxes can be labeled something similar to "limited," "intermediate," and "advanced."

If you prefer, you can make the assessment system a little more complex. Instead of having checkboxes marked limited, intermediate, and advanced, you can establish a 1 to 10 ranking system, with 1 be-

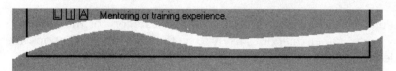

Job Category Requirements
Advanced Documentation Tester

Professional Behavior:

☐ ☐ Ⓐ Promptness.

☐ ☐ Ⓐ Works well under pressure.

Communication Skills:

☐ ☐ Ⓐ Ability to speak in meetings.

☐ ☐ Ⓐ Ability to make formal presentations.

☐ ☐ Ⓐ Ability to communicate with peers and superiors.

☐ ☐ Ⓐ Mentoring or training experience.

9-5 *Checkbox method of assessing skill level*

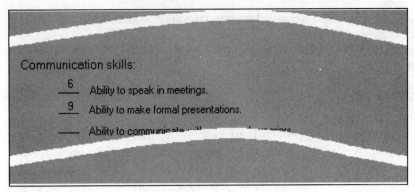

Communication skills:

__6__ Ability to speak in meetings.

__9__ Ability to make formal presentations.

_____ Ability to communicate ....

9-6 *Example of using the ranking method of assessing skill level*

ing no experience and 10 being expert-level experience (Fig. 9-6). There are advantages to both approaches.

With the first method, once you have filled out the assessment checklist for each employee, it is quickly obvious which skills need to be improved. For example, if the employee's skill in the area of making formal presentations is marked as limited, this might be the first skill that should be improved. If you have eight items marked limited, however, you will have to prioritize those eight items and work on the most important one first.

With the second option, prioritization and level of importance are already established. You might have more than one skill marked as a five or a four, but for the most part, there might only be a couple of each from which to choose.

Once you have created the skills assessment checklist, you then meet with each documentation tester. Together you determine the level of skill each tester has in each of the items on the list.

One of the most effective methods for determining a tester's level of skill is to go over each item on the checklist with the tester. Have the tester tell you where they believe their level of skill to be for each item. To ensure the tester does not overstate or understate his skill level, explain to him the purpose of the test—to help them improve their skills; not to determine whether they can keep their present position.

Once you and the testers have filled out a checklist for each tester, you can then move on to the next step—planning each documentation tester's training.

Plan the training

Once you identify the areas in which each of your documentation testers need training, you can begin planning for that training. There are two common approaches you can take to plan training for your documentation testers.

The first approach is a three-step process that concentrates on the needs of a single employee. To follow this approach, first choose the most important training issue for each employee. Next, determine the best approach to provide the needed training. Then finally, make it possible for the employee to get that training.

The second approach is to look at the training needs of all employees. Assess which needs are common to the majority of the testers. Determine the best approach to provide this training for each of the testers. Then arrange or make it possible for these employees to get that training.

The first approach is oriented toward the greatest need of each individual employee. The second approach is oriented toward meeting the needs that are most common to all documentation testers. The first approach is often the most expensive and the most time consuming. Because group discounts are available, the second approach is usually less costly. The method you choose for meeting a documentation tester's training needs also determines the cost of the training.

Regardless of the approach you choose, you should prepare individual training plans. As a matter of policy and procedure, training plans should be in writing and established for each employee, even if you will be implementing the training for several employees at a time.

Companies that use written job descriptions also generally have written plans for employee assessment, often referred to as the annual review process. If this is the case in your company, the training plans you write for a given employee can become part of the annual review. These plans are used to establish goals the employee is to meet in the coming year. Of course, the training goals are only a part of the total goals for the coming year.

The intent is to establish specific training goals that help the employee do a better job in his current position. These training goals might also help to prepare the employee for the next available promotion. As an added benefit, including these goals on the employee's annual review provides the employee with an incentive for fulfilling his training needs.

Even if the employee's training plans are not part of the annual review process, they should still be defined based on the employee's position requirements. For example, if one of the employee's position requirements is the ability to make formal presentations, yet the employee has ranked his skill as limited in this area, you can plan training to help him meet this goal.

When planning the training for your documentation testers, you must determine which approach you will take—concentrating on the individual tester's need, or finding and filling common needs. You must then decide the most suitable and cost-effective approach to fulfill those needs and implement the approach you have planned.

Regardless of the approach you take or the way you choose to implement the approach, there is one other thing you should do to help train your documentation testers. You should create a documentation tester training manual.

Create a documentation tester training manual

All companies have their own approach to performing different tasks and fulfilling specific job responsibilities. Even the most experienced documentation tester cannot be expected to know and understand a company's defined idiosyncrasies.

To help your employees learn those aspects of documentation testing in your company that might differ from other companies, provide your employees with a documentation tester training manual. Some companies refer to this type of manual as an employee orientation manual. Unless the only employees in the company are documentation testers, however, the employee orientation manual is not the same as the documentation testing manual.

The employee orientation manual generally contains information about working hours, business attire, vacation and sick leave policies, and other types of similar information. These policies apply equally to all company employees. The documentation testing manual, on the other hand, provides information particularly useful to the documentation tester.

The documentation testing manual contains a variety of information. If it is information that is useful to documentation testers, and not otherwise easily available, it should be included in the documentation testing manual. What the documentation testing manual contains in one company might be totally different from what it contains in another company.

For example, a software development firm that I worked for had a documentation testing manual. It contained, among other things, the following items of information:
- A copy of the process required to create documentation
- Samples of the forms documentation testers use to record software bugs and documentation errors

- Instructions for filling out the forms used to record software bugs and documentation errors
- Organization charts for the writing, testing, and product development departments
- A copy of a report prepared as the result of researching the preferred approach for writers and testers to follow when working with engineers
- A copy of the job description for each available documentation testing position
- The corporate president's letter stating the company's mission and purpose
- Sample testing reports

These items would be useful to documentation testers in any company. You can use this list as a starting point for creating your own documentation testing manual. Include in it anything you believe to be pertinent and useful to your documentation testers. When creating your manual, be careful not to include information in the manual that might violate company policy.

For example, some companies do not want employees to know the details of other positions available within their department. Or they might only allow employees to know information about the position considered to be directly below or above their current level. In this case, unless all of your documentation testers are currently employed at the same level, you probably should not include the different documentation tester job descriptions.

Other than taking these types of precautions, put anything into your documentation testing manual that you believe will be of benefit to your employees. If you are not sure where to begin, let me make a suggestion. Start by asking each of your current documentation testers to make a list of those things they discovered they needed to know but had to search to find out when they first started working for your company. Use the information on those lists as a reference for developing a table of contents against which you will then develop the manual's content.

Assign a mentor to new employees

Even when you develop a training plan for your employees, you still need to supply additional support for new employees. This is particularly true for the less-experienced testers you hire. Doing so has several benefits:

- Helps new employees become oriented to the company and their position as quickly as possible

- Helps to ensure they get questions answered with the least effort
- Makes them productive more quickly
- Gets them introduced to development and documentation team members in a quick and efficient manner
- Eases the transition of project assignments from one tester to another

One of the easiest ways to provide additional support is to assign one of the more experienced documentation testers to be a mentor to the new tester. Tell the mentor that you expect him to assist the new tester in any manner possible, so as to help the new employee be as productive as possible, as quickly as possible.

Besides benefiting the new tester, assigning a mentor benefits you as the manager as well. It reduces the amount of time you have to spend helping the new tester. In addition, the new tester becomes productive more quickly which might help the entire team work more quickly.

When assigning a mentor, consider personalities. Try to assign an individual who is most likely to get along with the new employee. You also have to consider which of your current employees can afford the time to spend with a new employee. If the employee you want to assign as a mentor has a workload that does not allow enough time to be effective as a mentor, you are defeating your purpose if you do not correct the situation. Either lighten the workload of the person you are assigning as the mentor, or consider using another individual as the mentor.

Give each tester the opportunity to be a mentor to new employees. If you always assign the same individual to be the mentor, you might be doing a disservice to that individual. Mentoring should be a small part of the documentation tester's responsibility. It should be an opportunity for them to learn and grow as well. It should not be a burden, and it should not interfere with their own personal growth.

Set long-term goals

The annual review, and the use of training goals as part of the annual review process, helps establish some of the documentation tester's short-term goals. Other short-term goals included in the annual review relate directly to other job duties, functions, and responsibilities. For the most part, any goals that are established with the intent of meeting those goals within one year or less can be considered short-term goals. Goals that do not need to be met within one year are long-term goals.

Short-term goals are easy to set. Define and set them based on the employee's job description and the company's project plans for the coming year. For example, if you know the company will be producing six new products over the next year, then you know the documentation for these projects will have to be tested. You can assign the testing of these projects as short-term goals to different testers, keeping in mind their skill level and product knowledge when making these assignments.

Long-term goals are more often related to special assignments than to daily job functions. For documentation testers, long-term goals often involve research into developing technologies or related items that might be of benefit to your company and the products it develops and markets.

For example, if your company develops and markets software, a long-term goal for a documentation tester might be to research and report on computerized methods for delivering documentation or for automating documentation testing procedures. Considering the long-term future of your company and the documentation tester's functions and responsibilities in that future is a tester's responsibility. As such, it should also be considered a long-term goal. Even though long-term goals are not assigned as frequently as short-term goals, when the option exists to assign a long-term goal, that option should be exercised.

Motivating the team

Now that you have a documentation testing team in place, and you have established short- and long-term goals and training plans for its members, you need to consider how to motivate that team. As discussed earlier, documentation testers are generally creative, energetic individuals, with a strong desire to learn. Because of their interaction and involvement with so many aspects of a company's products, they quickly become valued company assets. In a relatively short period of time, documentation testers gain a great deal of knowledge. When this happens, it becomes harder and harder for the tester to motivate himself. That is when motivation efforts aimed at individual testers, and the testing team as a whole, become particularly important.

There are various approaches you can take to help motivate the members of your documentation testing team. The six approaches discussed here are just a drop in the bucket. Volumes of information on motivation techniques are available and can be found with little effort. The six approaches discussed here are included because it is easy to overlook them as motivators:

- Managing employee stress
- Developing a service attitude
- Providing growth opportunities
- Defining objectives and expectations
- Developing employee action plans
- Monitoring employee performance

Manage employee stress

Almost every job carries with it a certain amount of stress. Being responsible for the quality of one or more pieces of documentation is no exception. The level of stress a documentation tester experiences depends on several factors. Some of these, such as the amount of work required of a tester, are not as easy to control as others. Whenever and wherever possible, you should try to help reduce an employee's stress.

One way to manage employee stress is to find out what the employee considers to be stressful. Ask your documentation testers which aspects of their job they find most stressful. Then, to the best of your ability, find ways to eliminate or reduce some of these stress factors.

You do not have to take on the entire responsibility of reducing stress by yourself. Employees have some responsibility for this as well, as do other departments within the company, such as human resources. Therefore, ask the employees as well as other company personnel to suggest or help you find ways to reduce employee stress. Choose those ways which seem likely to be the most effective, then make every effort to implement those stress reduction techniques you have chosen.

Just the fact that you care enough to help the employee control this negative aspect of the job helps to some extent to relieve some of the employee's job-related stress. It is important to recognize the potential causes and problems associated with stress and to try, for the sake of the employee as well as the company, to eliminate or lessen this stress.

Develop a service attitude

Both the documentation testers and the documentation testing team manager should work toward developing a service attitude. That is, both should believe that the job they do is more than just that of ensuring quality. It is a job of service. The job they do is a service to the company, to the team, to the product's users, to other employees, and to each other. Working toward accomplishing goals that result in the positive support and service of others, seems more rewarding. Therefore, developing an attitude of service can be a motivating factor in and of itself.

The documentation testing team whose manager has a service attitude toward the testers helps to encourage them to be of service to others. The manager's primary responsibility for service is to remove and prevent all obstacles from negatively impacting the team and its production. The tester's primary responsibility for service is to remove and prevent as many documentation-related problems as possible.

Maintaining this attitude of service, and accomplishing it, gives everyone involved a certain sense of satisfaction. Knowing you have the opportunity to do the best job you can, and that once you have you have provided a valuable service to others, can be somewhat motivational.

Provide growth opportunities

The need to provide documentation testers with opportunities to grow and develop was discussed earlier. As chapter 8 points out:

[Y]ou must make it possible for individuals to progress through the different levels. People want to be recognized and rewarded for their efforts. If you do not make it possible for your documentation testers to be rewarded with advancement opportunities, you will lose them to other positions or other companies.

Providing the opportunity for testers to advance in their positions is particularly important in the field of documentation testing. As you might have already figured out, it does not take long for your documentation testers to become some of the most valued employees in the company. Documentation testers soon gain more knowledge about the company's product than almost any other employee. They learn the product in intimate detail, function well in team settings, write effectively, and are self-starting and highly motivated individuals. Within two years, sometimes less, a documentation tester is generally ready to move up or on, and will find a wealth of opportunities outside your department if you do not provide them within your department.

Providing growth opportunities for your documentation testers, and showing them how to prepare for and take advantage of these opportunities, is a strong motivator. You have the opportunity to use the established company guidelines as more than just a function of the job. You also have the opportunity to use them as an employee motivator.

Define objectives and expectations

People like to know what is expected of them, including what they are expected to accomplish. Not only does that establish a clearly defined set of requirements, but it also provides motivation for the employee.

Having goals or objectives toward which one can work is often very motivational and can contribute to an increase in productivity. There are different reasons why this is so. Perhaps people believe that the only way to keep their job is to meet all of the objectives and expectations defined for them. Or perhaps there is a related incentive.

For example, one of the companies I worked for had a performance review system in place that was designed to reward esmployees for their achievements. The system required quarterly reviews and an annual review.

At each quarterly review, the employee and the employee's manager reviewed the employee's performance for the past quarter. Each of the assignments or goals set for that employee were compared with what the employee accomplished. Then new goals were set or the current goals extended for the coming quarter.

At the end of the year, the employee's accomplishments for the year were compared with the employee's original goals. Then the employee's goals were ranked by their importance to the job position, and the employee's accomplishments were scored with a ranking system of zero to four.

In this ranking system, a score of zero meant that the employee had failed to satisfactorily meet that goal. A score of four meant the employee had exceeded what was expected of them for that goal. The intermediate scores of one and three reflected a level between completely unacceptable (level zero) and acceptable (level two), and between (level four) greatly exceeding expectations and acceptable or meeting expectations (level two). All scores received for each category of job responsibility were totaled at the end of the year, and divided by the number of categories to determine the final ranking or score (zero to five inclusive), with fractions of points being rounded either up or down as tradition required.

Based on the final ranking or score, if the employee ranked in the mid-level (two), the employee could expect to receive a raise based on the average percentage that would be given to all employees qualifying for a raise. (You had to get a ranking of two in order to get any raise.) If the employee received a three or a four, the employee's raise percentage was adjusted upward to reflect the better-than-average job performance of the previous year. If your ranking was one, you could expect to meet with your manager to establish a plan for improving your performance. And a ranking of zero often meant you would never again be meeting with any manager in this company.

How much of a raise you received for the coming year was determined by how well you had done in the past year. The determination of the ranking on each job goal is somewhat subjective. However, the final total is not quite as subjective as it would be if the manager was simply determining the score based on a feeling. Instead, the final score is determined by a mathematical calculation.

This system, although not perfect, did present employees with the opportunity of being rewarded for hard work. By working hard in any given year, the employee could reap the financial reward in the coming year.

This system had one other benefit as well. The company also rewarded its employees with annual bonuses, if the company had a financially successful year. In those years in which a bonus was granted, the bonus that the employees received was based on a percentage of their annual salary. The more your salary increased during the year, the larger your bonus. But the big benefit was directly related to your score on your annual review. Those employees whose score ranked a three or four received a higher percentage bonus than did those employees who ranked a one or two.

By defining objectives or goals for each employee, and clearly spelling out what the employee must do to meet those objectives or goals, the employee was given a certain amount of control over their own working environment. In addition, those employees who chose to go all out and exceed the defined expectations were rewarded with salary increases and larger bonuses. Those who chose to meet but not exceed their defined goals were also rewarded with increased salaries and bonuses, but their increase and bonus was not as great as those who chose to work harder.

Employees had the flexibility to choose. They could exceed the minimum requirements that were clearly spelled out for them, and be increasingly rewarded because of it, or they could simply meet those requirements, retain their jobs, and be rewarded with reasonable salary increases and bonuses. This system provided incentive and motivation for each employee to accept and work with as it best fit their particular needs. It was as if each employee had their own personal incentive program designed just for them. This was the direct result of having had objectives and expectations clearly defined for them, and then rewarding them when they met or exceeded those objectives and expectations.

Of course, while many companies define objectives and expectations, not all companies use a reward system to encourage employees to meet those objectives and expectations. Some companies choose instead to use fear.

Vince Lombardi, a coach for the Green Bay Packers, knew how to use fear as a motivating tool. His team's objective was to win games, of course. To provide inspiration and help his team accomplish that goal, Mr. Lombardi, like many coaches, would give his team a pregame speech. One of his more brief but inspirational pregame speeches has since been quoted as including a single but important sentence. He told his players that, "If you aren't fired with enthusiasm—you will be fired with enthusiasm" (Griffith 1990).

Of course, reward and fear are only two examples of the motivational application of goals and objectives. Here is another example that is quite creative because it shows how people can be motivated by basic needs to accomplish clearly defined objectives (Griffith 1990).

The loaded station wagon pulled into the only remaining campsite. Four youngsters leaped from the vehicle and began feverishly unloading gear and setting up a tent. The boys then rushed off to gather firewood, while the girls and their mother set up the camp stove and cooking utensils.

A nearby camper marveled to the youngsters' father, "That, sir, is some display of teamwork."

The father replied, "I have a system. No one goes to the bathroom until the camp is set up."

Another basic need that can be both a motivation to and a reward for meeting objectives and expectations is team interaction. Even though documentation testing requires a great deal of "quiet" time in which to work, interaction with other team members is important in keeping employees motivated. Chitchat in the break room, phone calls, people stopping by your office to ask questions or see how you are doing all contribute to an employee's sense of well being, and thus to his ability to meet the objectives and expectations that have been set for him (Wallace 1994).

Psychologist and anthropologists have always known that if you take the man out of society, you risk taking society out of the man— leaving him feeling empty, alienated, depressed, and ultimately, ineffective. Even such a strong-willed character as the Polar explorer Robert Byrd found the burdens of isolation nearly fatal. Working alone in a meteorological observation post, he found himself prey to hallucinations and morbid fantasies and became heedless of his health and safety.

While most jobs do not require such extreme isolation as that which Robert Byrd experienced, being closed up in your office every day, with limited or no interaction with others, is not the best way to ensure you accomplish your objectives. Instead, making time for team interaction, even though it takes some time out of your day, keeps you more mentally alert and enthusiastic about your work. Ultimately, the little bit of time "wasted" at the break room or in brief nonbusiness conversations might cause you to be more productive in the long run than constantly keeping your nose to the grindstone. A manager who recognizes this fact can encourage and use team interaction to motivate and encourage team members.

Defining objectives and expectations for your employees is the first step in using objectives and expectations to motivate your employees. Providing rewards or incentives, no matter how small or basic, even sometimes including a little fear, makes those defined objectives and expectations even more motivational.

Develop employee action plans

Objectives and expectations tell the employee what should and must be done. To meet those objectives and expectations, however, employees sometimes need a little guidance to show them how to take appropriate action and what action is appropriate. Employee action plans provide the details needed to help employees know what they are to do in order to meet the objectives and expectations defined for them.

For example, assume one of the objectives defined for each of your documentation testers is to share new knowledge and experiences with other team members, thus broadening the knowledge of all documentation testers. This is a fairly broad and generic objective—learn something and share it. To ensure that documentation testers can successfully meet that objective, you should give them some guidelines as to what they should do to meet it.

There are several options, of course. You might be able to think of some right away or you might need to research the options open to you. Some options might have you considering the purchase of a book relevant to your field (such as this one), which you would then read, and for which you would then prepare and present a 10- or 15-minute report at your next team meeting. This is a reasonably simple and inexpensive approach to meeting this objective.

On the other hand, you might choose to join a related society or organization, attend meetings and conferences, and report back on what you have learned. A slightly more time-consuming and costly approach to meeting the same objective.

In these two instances, the money and time involved vary greatly. But, both meet the generic objective of sharing new knowledge with other team members. However, one of these two approaches might be more acceptable to the company than the other. The employee needs to know what is and is not sufficient and appropriate for meeting the defined objective. This is where an employee action plan becomes important.

Employee action plans can be created by the employee's manager and passed on to the employee, or they can be created as a joint effort between the manager and the employee. The latter is preferable because part of the motivational aspect of employee action plans is the fact that the employee has some input into their content.

An employee action plan should contain a list of tasks to be completed in order to accomplish the objectives and expectations defined for each employee. As a minimum, the plan should list those tasks to be completed. When possible, list additional tasks to be completed that meet the objective or expectation, but which, if completed, are considered going beyond what is minimally expected of the employee. Such an approach provides the employee the opportunity and the specific direction to be taken if that employee chooses to do more than what is minimally expected. This approach is often a mutually beneficial situation, particularly if the employee is rewarded in some manner for his additional efforts on the company's behalf.

Monitor employee performance

One additional way to help motivate an employee is to monitor his performance. No one likes to be hovered over, checked on, or otherwise constantly watched. However, no one much likes being ignored either. In the business world, reaching a happy medium can be effective when trying to motivate employees.

Some ways in which employee performance can be monitored and feedback provided to that employee include

- Holding regular team meetings in which each employee has the opportunity to discuss what they are working on, how it is coming along, and any problems they have that need to be solved. Feedback, particularly positive reinforcement and help in solving any problems, is an effective motivational factor.
- Writing and delivering biweekly reports to management shows both the employee and the manager what has been accomplished during the previous two weeks. This is also the place to list what the employee intends to accomplish during

the coming two weeks. Breaking the tasks associated with objectives and expectations into smaller, more manageable units helps both the employee and the manager see some accomplishment, even for larger projects that are generally expected to take several weeks or months to complete.

- Conducting quarterly performance reviews in which the manager and employee discuss what was accomplished during the previous quarter, and then plan what is to be accomplished for the coming quarter can also be motivational. Holding quarterly performance reviews lets both the manager and employee know what is expected. In addition, the manager can take the opportunity to correct any misdirection, and recognize the employee for past accomplishments and successes.

- Rewarding an employee for a specific task or accomplishment is another way to motivate an employee. A common method used is an employee-of-the-month award. The employee's peers or manager are asked to nominate an employee and specify why that employee should receive the award. The award is given to the employee, usually at a meeting, so that everyone has the opportunity to recognize the employee with a word of congratulations or a hand shake. By recognizing an employee for a job well done, the employee feels that the company appreciates his extra effort. This motivates the employee to continue putting out that extra effort, and is generally a small price to pay for continued employee performance.

Other companies have other methods of motivating employees as well. As noted at the beginning of this section, the six specific topics discussed here are some of the least common or the most often over looked opportunities for motivation. You might want to consider which of these you can implement for your documentation testing team, as well as how you can successfully implement them. In addition, you can use these topics to help you think of other effective and creative methods of motivating your employees.

Summary

In an ideal world, every employee you hire as a documentation tester will be an experienced tester and possess an intimate knowledge of your product, its market, its users, and its competition. In addition, they will be excellent writers, editors, and speakers, and excel in user interface design, product development, and documentation development. They will also be willing to work long hours for the company's idea of reasonable pay, be self-motivated, always do more than is asked of them, always know what to do without being asked, and never have to be rewarded or shown appreciation for their efforts.

Well, we do not live in an ideal world, so there are no absolutely perfect employees. What you have to work with are, we both hope, decent people with reasonably normal working habits and ethics, who need some guidance and appreciation, in addition to sometimes needing additional training and motivation to help them to give you their best.

This chapter concentrated on teaching you what options are available for training and motivating your documentation testing team employees. It discussed how to satisfy your employee's training needs, suggesting that you take several specific steps.

The steps for training documentation testers began with the recommendation that you develop a skills assessment checklist to help you determine what training your employees need. That suggestion was followed by a recommendation that you plan the appropriate training, and create a documentation testing manual to fill in specific items of information each documentation tester needs to know which might not otherwise be provided through the planned training. Next, it is suggested that you assign a mentor to each new employee, and that you set long-term goals for each member of the documentation testing team. Finally, even while you are still training enthusiastic new documentation testers, this chapter suggested six approaches to motivating your team members. These approaches covered ways to manage employee stress, develop an attitude of service, provide growth opportunities for your testers, define objectives and expectations for each employee, develop employee action plans, and monitor employee performance.

Of course, there are many approaches you can take to motivate employees, but the six discussed in this chapter are some that can be easily overlooked. In addition, while considering these six approaches to motivating your employees, it is hoped that you might also be inspired to come up with some creative, unique, and particularly effective approaches of your own for motivating your employees.

Establishing a documentation testing team is a great deal of work. Once you do get the team established, you want to help them be the best and most productive employees they can possibly be. Ensuring they get the training they need, and providing effective motivation, helps them to be the best documentation testers they can be, and helps you build and grow an effective, productive, and successful documentation testing team.

References

Griffith, Joe. 1990. *Speaker's Library of Business Stories, Anecdotes and Humor.* Englewood Cliffs, New Jersey: Prentice Hall.

Wallace, Don. "Overcoming Isolation: How to Stay Connected When You're Self-Employed." *Home Office Computing* (March 1995): 59–65.

10

Some final guidelines

Anything worth having rarely comes easy. An old cliché, I know, but a true statement none-the-less. And starting, developing, and managing a successful, productive, and efficient documentation testing team is no exception. As you work toward developing a documentation testing team in your company, or even just as you work toward developing yourself as a successful documentation tester, you will have problems and setbacks. There will be days when you wish you had never started this whole thing. But the end result, better document and product quality, will one day be worth it all.

In the mean time, you must deal with daily realities. Those realities include making mistakes, getting off in the wrong direction, dealing with people who believe they know documentation testing better than you or your department does, and a host of related realities. This chapter is designed to bring to light some of those realities with which you must deal so that you can be better prepared for them when they come along.

After reading this chapter, you will understand:
- Common mistakes documentation testers and documentation testing team managers make and how to avoid them
- Ways to keep your documentation testing team on target
- What to do and consider when others try to dictate the documentation test to be conducted
- Why it is important to maximize each team member's special talents and abilities, and how to go about doing so
- Some approaches you can take to get recognition for your documentation testing team

Common mistakes
and how to avoid them

Several mistakes are common to documentation testing:

- Underestimating documentation testing
- Expecting too much of documentation testing
- Assuming that just anyone can do documentation testing
- Promising to do more than you can do in the time allowed
- Promising to deliver the tested document in a time frame that is too short for the type of test to be conducted
- Forgetting that your job is a service you are providing to and for the document's writer(s) and users

The most common mistake new testers and testing team managers make is that of underestimating the power of documentation testing to significantly improve the quality of product documentation and its associated product. As power brings with it responsibility, underestimating the power means you might also underestimate the responsibility. This mistake can manifest itself in several ways:

- Being too timid when representing documentation testing
- Expecting people to respond negatively to the possibilities of documentation testing
- Assuming the documentation testing team has no power to influence the direction of documentation and product development
- Relinquishing documentation testing-related decisions to individuals without sufficient experience and knowledge to make effective decisions related to documentation testing

I hope you are beginning to see the point. Documentation testing is a relatively unknown field, and as such, it is easy to be mislead into believing that it has only limited value and merit. Thus, not giving it the respect it deserves is a common mistake. The opposite is also true.

If you assume that documentation testing is the answer to all of your company's quality problems, you have also made a mistake. Documentation testing is not a miracle pill or a cure-all. It is a method, perhaps even an art or a science, which, when properly followed, can substantially improve your product's documentation. In the process, it might also help to improve your company's product. It cannot, however, solve other problems your company might have. Documentation testing's goal, function, purpose, and result is improved document quality. That is what you must keep in mind, and what you must work toward.

You should neither think more of documentation testing than it is, nor should you think less. Look at documentation testing with a realistic attitude, approach it from a knowledgeable standpoint, then use it for all it is worth, but know its limit. If you do so, not only will your company benefit, but so too will you.

Another common mistake that both individuals and companies make when it comes to documentation testing is to assume that anyone can do it, thus assigning just anyone to do it. To successfully affect the quality of the documentation your company produces requires a certain minimum set of skills and personality traits. This book has discussed a wide range of both.

Skills a documentation tester needs to be successful include such things as organizational skills—the ability to meet deadlines, the habit of being a self-starter, the desire to learn, and a certain skill at and satisfaction in problem solving, as well as a variety of other skills, including

- Product experience
- Writing skills
- Editing skills
- Business sense
- Industry knowledge
- Computer aptitude

Some of the personality traits that help to make for an effective and successful documentation tester are

- Group participation skills
- Communication skills
- Product aptitude
- Ability to handle stress
- Ability to adapt to change
- Organizational skills

As time goes on you might be able to add to these lists of skills and personality traits. In addition, you will likely find that some skills and personality traits are more important than others, and that in a pinch, employees who have a respectable combination of the more important ones can become effective documentation testers. But for the most part, using whoever is available at the time to fill a documentation tester is often unsuccessful or can even worsen the problem.

Using someone who is unqualified to be a documentation tester can damage the reputation of the documentation testing team as a whole, and reduce instead of improve the quality of the documentation. In addition, taking an employee who is capable, successful, and qualified in their present position, and moving them into a position they neither enjoy nor are qualified for, is likely to result in the loss of the employee.

Therefore, carefully screen, choose, and train your documentation testers. Make certain they add value to the company and its documentation. Reduce as much as possible any negative impact the establishment of a documentation testing team might initially have on the company. To accomplish this, ensure that the documentation testing team is made up of the best-qualified, most-capable individuals.

As a documentation tester, the three common mistakes you might make but which you should try to avoid are the last three in the list near the beginning of this chapter:

- Promising to do more than you can do in the time allowed
- Promising to deliver the tested document in a time frame that is too short for the type of test to be conducted
- Forgetting that your job is a service you are providing to and for the document's writer(s) and users

No matter how hard you try, you can only do so much in the time allotted. In order to do more, you must greatly extend your working hours, which endangers the quality of the service, or you must get concessions from those demanding a quick turn-around time regarding the level of test you will conduct or the quantity of documentation you are to test.

As a new tester, you probably have a desire to satisfy everyone and make all parties happy. While that is a commendable attitude, it is not very practical. Documentation testing is a very demanding profession, both in time and effort. Even experienced testers sometimes find it impossible to meet the schedules they have established. There are many reasons for this, but for an experienced tester, the reason is not usually because they committed to do too much in too short a period. However, that is often the case with new documentation testers.

To prevent this from happening to you, try not to give any time or testing level commitments until you have had a more experienced tester review the project with you. If possible, have your assigned mentor, manager, team leader, or a more senior tester look over the requirements and time schedule for your project and provide you with some input before you make any commitments.

If you are put on the spot and not given an opportunity to seek assistance from a more experienced tester before providing a time or level estimate, then tell the requesting party that you can only give them a rough guess, and that it is subject to change. Then guesstimate how much time you think it will take you, and double that estimate. As to the level of test, tell the requester you need to look at the document a little further before you can provide that information, but that you will do so as quickly as possible.

Do your best to get out of the situation graciously. Then make every effort to quickly determine a more accurate time frame and testing level, seeking out whatever help you need in order to determine this information. Once you have done so, get back to the requester with the information as quickly as possible.

Whatever you do, when pressed for the requested information, remember you are there to provide a service that benefits everyone who might be affected by the service. Also remember that most of the other company employees are there for the same reason. Try to keep your associations amicable, and keep a professional attitude and approach.

Keep your team on target

As a team leader or manager for a documentation testing team, it is your responsibility to direct the team. When your company is one in which multiple projects are being simultaneously developed, or the quantity of documentation being produced seems like it is too much to keep up with, it is your job to keep your team on target and on track.

Two factors can derail your team and keep them from reaching their ultimate goal—high-quality documentation, on time, and within budget. The first factor has to do with the team itself. The second factor relates to the company and its other employees.

Teams, in order to function at their peak efficiency, must be cohesive. That is, the team must feel and work like a team, rather than like a group of separate individuals all doing the same type of work. In addition, the team must feel good about what they do, and each member should feel appreciated for his efforts and team involvement.

One employee with morale problems or who has not learned how to function as a team member can quickly destroy the cohesiveness of the entire team. Therefore, it is important to ensure that each team member is happy, productive, and successful in their job. If you have a team member who is not, make every effort to correct the problem within the constraints of company policy. If you cannot correct the problem, remove the problem, or the entire team and the company will suffer as a result.

Just as you must address problems created within the documentation testing team, you must also address problems that come from outside the team. There are two categories of problems you will commonly deal with—people who "attack" the team, its members, its function and purpose, and a lack of recognition.

The biggest reason documentation testing and other company teams suffer an "attack" from within other areas of the company is politics. I will not even try to define the term. Suffice it to say that all companies have employees who play their little games. Most of the time, those games involve power—gaining it, keeping it, and wielding it.

As a documentation testing team manager, it is your responsibility to watch for these politically motivated attacks and to insulate your team members from them. While I cannot tell you specifically how to do that, I can offer some general advice.

If you see an attack coming, try to head it off by speaking directly with the person who is spearheading the attack. Or if it would be more beneficial, speak instead with that individual's superior. Bear in mind, however, that your approach might be viewed as an attempt to go over the other individual's head or to cause trouble.

If an attack is already under way, keep detailed records with which to defend yourself and your team. In addition, make an effort to find out the real reason for the attack. If possible, attempt to diffuse the situation. But use productive and professional tactics when doing so.

If the battle is over and you have already lost, try to make the best of the situation and make every effort to protect your team. Then, assuming you are still with the company and have the opportunity to do so, rebuild credibility, effectiveness, and anything else that was destroyed or damaged in the upheaval.

Prevent others from dictating testing

Other people in the company might have a vested interest in the type of documentation tests you conduct, how long you are allowed to conduct them, how active you are on documentation and development teams, and other aspect of your role as a documentation tester. Try to remember that the responsibility and the job is yours. Do the best you can in all situations.

If a writer insists that you perform a specific type of test, for example, you have a right to be skeptical. Ask the writer to explain why the level or type of test being requested would be the best test to perform. If the writer has valid reasons, such as the document received a thorough test last time around and only minor changes have been made since then, you can comfortably acquiesce to the writer's request.

On the other hand, if the writer cannot give you a good reason, you must consider the possibility that the document is not what the writer hoped it would be, and that the writer would rather no one found out. If you believe this to be the case, it is your job to let the writer know you will make an effort to comply with the request, but that your goal is to help the writer to produce the best possible document within the current restraints.

It might be necessary for you to remind the writer that you are not there to criticize or belittle, and that you understand the constraints within which you all must work. You might also choose to remind the writer that sooner or later, problems that you do not catch and fix in the document will be discovered, and then both the writer's job and your job will be on the line. In other words, support the writer's emotional response to the possibility of someone discovering they are not perfect by letting the writer know that you know your job is to make both the writer and the document look and be better.

Of course, there might be others who try to dictate the type of test to be conducted, or the start and end times and test length. You have to resist the temptation to cave in and do whatever you are asked, even if you know it is wrong.

If the dictate regarding the test comes from a superior in your department, cave in. There is no point in arguing with someone who is knowledgeable about documentation testing as well as the dictates of upper management. (If you do, you might be seen as one of those testers who cannot get along with people or who disrupts the cohesiveness of the team.) But if you are absolutely convinced that conducting the testing as instructed will be harmful to the company, its product, or its product's users, you must bring this issue up. Do so tactfully and respectfully. Once you have been heard, follow management's decision.

If the dictate regarding the test comes from outside your department, take it to those within your department who can effectively deal with it, if you cannot do so yourself. There might be occasions when you cannot effectively or politically deal with a situation. If the dictate comes from management, it might be up to your department's manager to deal with it. You have to consider the situation and the source and make a decision that you deem is best for the product documentation and the company, then go from there.

There is one other external influence that might affect the documentation team. That is the lack of recognition from other areas within the company. The documentation testing team knows they add value to the product's documentation, and often to the product itself. But it can be very frustrating if it seems as though no one else knows it.

If it is true that no one else knows it, the documentation testing team might frequently be listed among the "okay to cut" items when budget reduction time rolls around. To prevent this from being the case, and to see to it that others know and understand the value of documentation testing and the value a documentation tester adds, you have to be the testing team's champion (also known as a defender of evil, promoter of the cause, and so on). The last section of this chapter provides some general recommendations for championing the documentation testing team and its members.

Maximize team member talents and abilities

Some of the value a documentation testing team gives to a company is the wealth of its diversity. Testers come to a company with a variety of experiences, skills, and knowledge. Once there, they generally continue to grow in the position of documentation tester. In fact, some team members become so knowledgeable about a product that the team member is seen as too valuable to reassign. This team member is then left to slowly stagnate and become bored. Avoid this situation if at all possible.

Maximize your individual team members talents and abilities. This means you should use them on a variety of projects. Doing so not only benefits the team member and helps him to grow, but provides cross-training for your other documentation testers as well.

A team of testers who are cross-trained in a variety of company products provides a pool of talent from which to choose when new projects come along. In addition, it makes it possible for one tester to help another when the workload becomes to intense. It also makes it possible for one tester to pick up where another left off, in the event the first tester cannot finish a project for one reason or another.

Utilizing all of the talents, skills, and knowledge of each documentation tester makes the sum of the whole stronger and more valuable than the sum of its individual parts. It adds value to the company, provides variety to the team members, and helps each team member to grow and feel valued.

Get team recognition

One other factor that helps each team member feel valued is that of respect and recognition from other company employees. The only way to get that, because documentation testing is only now becoming more widely known and appreciated, is to ensure company employees know all about the documentation testing team. In other words, you must promote documentation testing and the testing team within your company, almost as actively as marketing and sales departments promote your company's products. To help promote the documentation testing team and its members, consider the following approaches (Fig. 10-1).

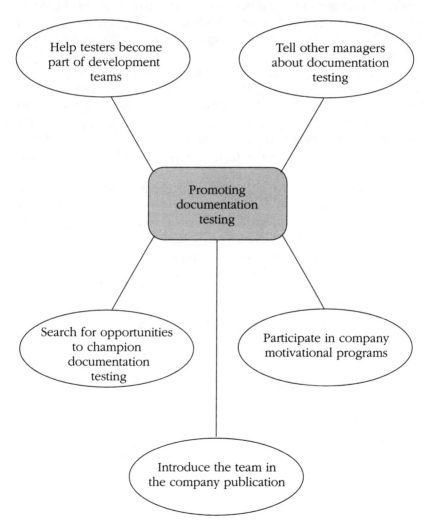

10-1 *Approaches to promoting documentation testing within your company*

Whenever possible, get documentation testing team members accepted as members of product development teams. At the first meeting the tester attends, have the tester briefly explain the function and purpose of documentation testing and how this adds value to the product and its documentation.

Tell other managers about the team. If you are allowed to do so, and your company conducts management team meetings, present the managers with a brief introduction and overview of documentation testing. Concentrate on the benefits documentation testing brings to the company. Dispel any rumors the managers might have heard. Try to show the managers how beneficial the team will be to all of their efforts.

If your company practices motivational techniques, such as giving out employee-of-the-month awards, be certain to nominate your best tester for that award. Write a brief description of why this tester should be chosen, and if possible, slip in a brief description of the purpose and function of documentation testing. Be sure to read your brief description in front of as many company employees as possible when the award is granted.

If your company has a company publication, make an effort to have the documentation testing team introduced in it. Help the publisher and writers to understand what documentation testing is, how it benefits the company, and why it is a good idea for them to write an article about the team. Then, of course, inform the team of what you have done, why you have done it, and when they can expect to be interviewed and maybe even have their picture taken.

In looking around your company you will see many opportunities to champion documentation testing. (You might even buy multiple copies of this book and slip them anonymously on people's desks.) Whenever an opportunity presents itself, take advantage of it. Let people know how important documentation testing is to the quality of the company's product documentation. Help them to see how important the documentation testing team's members are to the company and its products.

As time passes, your department and its employees will become recognized for their contribution. Then, unfortunately, you will probably have to keep training new employees to replace those who have gone on to other positions in your company. There is one high point here, however. Not only have you helped the company and these individual employees, but you have begun to spread the word about documentation testing. The more ex-testers you have in other company departments, the more documentation testing champions (by example and by word) there will be out there. Your championing activities will become less necessary, and eventually unnecessary altogether.

Summary

It is often said that anything worth having rarely comes easy. Some effort is required for all successes, and some problems and battles are inevitable. This chapter was designed to help you understand some of the problems and battles you will have to face if you are going to champion the cause of documentation testing in your company.

This chapter concentrated on explaining some of the common mistakes documentation testers and documentation testing team managers make, and on providing information to help you avoid these mistakes. In addition, this chapter attempted to provide you with useful information about ways to keep your documentation testing team on target, what to do and consider when others try to dictate the documentation test to be conducted, why it is important to and how to go about maximizing each team member's special talents and abilities, and some approaches you can take to get recognition for your documentation testing team.

Some of the information this chapter provided was aimed primarily at individuals who have chosen to become documentation testing managers, and to champion the cause of documentation testing. Some of the information was aimed at helping individual testers avoid the most common problems and pitfalls associated with documentation testing. Whichever you are or want to be—manager or tester—this chapter and this book should give you some insight into documentation testing as a functioning entity within a company.

With the knowledge contained within the pages of this book, you can be on your way to planning for, training for, developing, and managing a documentation testing team in your company. Or you can be on your way to becoming a documentation tester. Whichever you choose, I hope you will keep this book as a reference and source of inspiration, and that it will bring you success.

Glossary

abend Abnormal end. Often accompanied by a message issued by a software operating system when it detects a serious problem such as a hardware or software failure.

acronym A word created by using the first letters of a term or group of words.

alignment A group of people functioning as a single unit in order to reduce energy waste, establish a common direction, ensure synergy, share a single vision, and ensure an individual's singular efforts complement the unit's goal.

alpha test Having a select group of individuals work with a product before it is finished in order to seek feedback and implement suggestions prior to having intended users work with the product.

audience People (the reader) for whom a document is designed and written.

beta test Having intended users work with a product before it is finished in order to seek feedback and implement suggestions. This is usually done as a final test before a product is released. It is performed after changes have been made to a product as the result of an alpha test.

brainstorming The free exchange of ideas without fear of reprisal, reprimand, or insult.

budget An itemized summary of probable expenditures and income for a given period; a systematic plan for meeting expenses in a given period; the total sum of money allocated for a particular purpose or time period.

CBT Computer-based training. Electronic training programs developed to educate or train.

cell Point at which a vertical column and a horizontal row intersect, most commonly referred to in accounting spreadsheets.

DOS Disk operating system. Software responsible for the basic operation of a specific type of microcomputer.

end user Those who ultimately purchase and use a product.

303

FCS First customer ship. The date on which the final product is to be delivered into the hands of the first customer.

hardware Electronic equipment associated with a computer and any equipment that can be attached to it such as a printer.

holes Areas in documentation where needed information is missing.

ISBN Internal standard book number. Numbers assigned to books to ensure each has a unique number by which it can be identified.

MRD Marketing requirements document. A document describing the product's overall design from the marketing point of view.

opportunity cost The sacrifices made by foregoing benefits or returns as the result of choosing one of several alternatives over the others.

PC Personal computer.

PDR Product design requirements. The document that provides information about the requirements or needs the product will be designed and developed to fulfill.

prototype An example of a product from which to further develop and refine that product.

scope The depth of the project; often defined as a list of requirements or needs the project is to meet.

software Programs used by computers to provide instructions from which they perform various tasks.

spreadsheet A table of information contained within cells (points at which vertical columns and horizontal rows intersect).

target audience Those who are expected to purchase and use a product and its associated documentation.

test A Latin word (testum) meaning an earthen pot or vessel. Pots of this nature were originally used to assay metals.

TQM Total quality management. A system for ensuring quality throughout the design, development, production, delivery, and support of a product.

usability Capable of being used, in a fit condition for use, or intact or operative.

usability testing A research tool designed to find and correct deficiencies in products—primarily computer-based and other electronic products—and the documentation that accompanies those products.

user friendly Easy for the intended purchaser to understand and work with.

user interface The portion of the product with which the user interacts.

WAC Women's Army Corp.

WET team Writer/editor/tester team. The team of people responsible for developing a product's documentation. They are responsible for ensuring that the documentation meets the needs and wants of its audience and target customer, and that it is complete, accurate, and of the highest possible quality given the development time frame and dependencies.

winning publication A document that has been carefully planned, well written, precisely edited, gloriously illustrated, beautifully composed, and printed with care.

Index

Illustration page numbers are in **boldface**.

About the author

A former award-winning senior technical writer for Novell, Inc., and a former team leader for the Technical Publications Documentation Testing Team at Novell, Dorothy Cady is currently a freelance author and networking instructor. She is the author of *Inside Personal NetWare, CNA Study Guide*, and *New Rider's Guide to NetWare Certification*; coauthor of the *NetWare Training Guide: Managing NetWare Systems, 2nd Edition*; and a contributing author of *NetWare Training Guide: Networking Technology, 2nd Edition*. Dorothy Cady is also a Certified Novell Engineer (CNE), Certified Novell Instructor (CNI), Certified Novell Administrator (CNA), and an Enterprise Certified Novell Engineer (ECNE).